Handbook of Synthetic Organic Chemistry

Handbook of Synthetic Organic Chemistry

Second Edition

Michael C. Pirrung
Department of Chemistry, University of California, Riverside, CA

ELSEVIER

AMSTERDAM • BOSTON • HEIDELBERG • LONDON
NEW YORK • OXFORD • PARIS • SAN DIEGO
SAN FRANCISCO • SINGAPORE • SYDNEY • TOKYO
Academic Press is an imprint of Elsevier

Academic Press is an imprint of Elsevier
125 London Wall, London EC2Y 5AS, United Kingdom
525 B Street, Suite 1800, San Diego, CA 92101-4495, United States
50 Hampshire Street, 5th Floor, Cambridge, MA 02139, United States
The Boulevard, Langford Lane, Kidlington, Oxford OX5 1GB, United Kingdom

Notices
Knowledge and best practice in this field are constantly changing. As new research and experience broaden our
understanding, changes in research methods, professional practices, or medical treatment may become necessary.

Practitioners and researchers must always rely on their own experience and knowledge in evaluating and using any
information, methods, compounds, or experiments described herein. In using such information or methods they should be
mindful of their own safety and the safety of others, including parties for whom they have a professional responsibility.

To the fullest extent of the law, neither the Publisher nor the authors, contributors, or editors, assume any liability for
any injury and/or damage to persons or property as a matter of products liability, negligence or otherwise, or from any
use or operation of any methods, products, instructions, or ideas contained in the material herein.

Library of Congress Cataloging-in-Publication Data
A catalog record for this book is available from the Library of Congress

British Library Cataloguing-in-Publication Data
A catalogue record for this book is available from the British Library

ISBN: 978-0-12-809581-2

For information on all Academic Press publications
visit our website at https://www.elsevier.com

  Working together
to grow libraries in
ELSEVIER Book Aid developing countries
International

www.elsevier.com • www.bookaid.org

Publisher: John Fedor
Acquisition Editor: Katey Birtcher
Editorial Project Manager: Jill Cetel
Production Project Manager: Paul Prasad Chandramohan
Designer: Maria Ines Cruz

Typeset by TNQ Books and Journals

Disclaimer for Figure 8.3: Elsevier is independent of and disclaims any association with PerkinElmer or its affiliates. Any
screen shots taken from the Elements® SaaS solution include copyrighted material of PerkinElmer and its affiliates.

The content in this book is intended for instructional and informational purposes. To the extent permitted under
applicable law, neither Elsevier nor its suppliers and licensors assume responsibility for any injury and/or damage to
persons, animals or property as a matter of products liability, malpractice, failure to warn, negligence or otherwise, or
from any use or operation of any ideas, instructions, methods, tests, products or procedures displayed in the Content.
Practitioners and researchers must rely on their own experience, knowledge and judgment in evaluating or applying
any information, which remains their professional responsibility. Discussions, views, and recommendations expressed
may not be considered absolute and universal for every situation. Elsevier will not be liable for the failure by any user
of the Content to use due care in the use and validation of results made available through the Content.

Contents

Foreword

Efficient laboratory research in synthetic organic chemistry requires a remarkable number of skills. Day-to-day decision making is at the forefront, as the experienced experimentalist is rarely carrying out exactly the same transformation from one day to the next. This unique book, *Handbook of Synthetic Organic Chemistry, Second Edition*, provides a step-by-step guide to carrying out research in this challenging area. As noted by the author in the Preface of the first edition, the aim is to guide the "novice chemist making the transition from organic teaching laboratories to the synthetic chemistry research laboratory." In addition to admirably accomplishing this objective, this book brings together in one place wealth of information, which experienced researchers will also find useful.

As the first edition, this book is organized in a chronological fashion to provide the researcher with practical information from initially planning an experiment, to carrying it out, isolating products, cleaning up after the reaction, and determining the structure of products. This second edition incorporates much new information, beginning with an inaugural chapter on safety and extensive appendices on safety protocols. Considerations in selecting green solvents; procedures for handling chemical wastes and disposing hazardous reagents; and curated references to videos, softwares, and smartphone apps are now incorporated. In addition, many sections have been extensively revised and augmented. For example, the discussion of carrying out reactions using microwave heating now covers several pages and includes a useful table of the capacity of common solvents for microwave heating.

I extend my congratulations to Michael Pirrung for assembling such a useful practical guide to the practice of synthetic organic chemistry. Advanced undergraduates, graduate students, and practicing synthetic organic chemists are certain to find much helpful information in this extensively revised second edition.

Larry E. Overman
July 2016

Preface to the First Edition

I hope this book will be a useful indoctrination for novice chemists making the transition from organic teaching laboratories to the synthetic chemistry research laboratory, either in academe or industry. I also attempted to assemble some of the more useful but hard to locate information that the practicing synthetic chemist needs on a day-to-day basis. My aspiration for this book is to find it (with several tabbed pages) on chemists' lab benches. Finally, I aim to remind all readers of the little details about lab work that we may learn at some point in our careers but easily forget. When you are vexed by a particularly challenging experiment, I hope that paging through this book is one approach you take to solving your problem of the day, and that it is concise enough to encourage you to do so.

I organized the book to parallel the processes involved in planning, executing, and analyzing the synthetic preparation of a target molecule. I included a new chapter not found in earlier books on this subject matter: an example of the different formats in which the synthesis of a known compound may be published. I hope this chapter assists novice chemists in translating experimental descriptions into action items for today's experiment. I also found on the Web many new and valuable electronic resources contributed by the community of synthetic chemists.

This book has been over 25 years in the making. I first learned of an effort to assist beginning experimental students in learning the ropes of research laboratory work while a postdoctoral fellow at Columbia in 1980. Clark Still was giving a minicourse to his students on how to work in the lab. This seemed a very worthwhile activity to me, knowing how inept I was in the lab at the beginning of my graduate career. That pile of handwritten notes from Still's lectures eventually grew into a typed document that was finally scanned into electronic form. Along the way, it was distributed to my graduate students and postdocs in whatever its then-current state. Lately I have searched in earnest for books with comparable content that were comprehensive and modern, and was unable to find both in one text. However, I acknowledge my debt to those who have made past attempts at this sort of synthetic chemistry boot camp. I was lucky to be able to persuade Darla Henderson that this subject would be useful and popular, and it developed into the book presented here. I initially envisioned it would be titled *The Novice's Guide…*, but the opportunity she offered to echo the iconic *Chemist's Companion* penned by Gordon and Ford proved irresistible. My effort is offered in admiration of their work, and not the presumption that I can meet their high standard. I also want to be sure to recommend *The Laboratory Companion* written by Gary Coyne. It is truly a comprehensive guide to the hardware of the research laboratory, though it does not really touch on the specialized "software" of synthetic chemistry.

Finally, to the novice embarking on the study of organic synthesis, let me give you this advice: Lasciate ogne speranza, voi ch'intrate. This is the inscription above the Gates of Hell in Dante's Inferno (in the 1882 Longfellow translation, "All hope abandon, ye who enter in!"). Or, to quote a modern poet, Willie Nelson: "It's a difficult game to learn, and then it gets harder," in this case referring to golf. Synthetic organic chemistry can be one of the most frustrating, maddening, and capricious of scientific endeavors. For just this reason, success in synthesis is one of the most rewarding experiences in science. Synthesis is an intrinsically creative activity, and a chemist who does it well is often also creative in another area, be it music or cooking. If you already partake in creative hobbies, such as woodworking or knitting, you can anticipate synthesis offering you similar rewards. The achievement of the total synthesis of a complex target molecule is a peak experience for synthetic chemists, often celebrated with champagne. Even the small, day-to-day successes in the synthesis lab provide a great feeling of accomplishment. Once these are experienced, I expect you will be hooked. Hopefully, this book will help your "addiction" be its most fruitful.

Michael C. Pirrung

Preface to the Second Edition

I greatly appreciate the opportunity afforded by Academic Press and editor Katey Birtcher to update here what was originally *The Synthetic Organic Chemist's Companion*. Most of us would like to have a "mulligan" for work we did earlier in our careers, but rarely do we get the chance.

There are quite a few additions and improvements to that earlier book in this *Handbook*. It includes greater coverage of chemical safety, which certainly has seen increased awareness in academic synthetic laboratories since the *Companion* was published. From the opening chapter on general safety principles to hazard class protocols in the Appendix, safety topics appear frequently. Throughout, safety note boxes address important safety issues concerning the topics at hand. On related matters, new sections discuss destroying hazardous reagents and handling chemical waste.

Expanded discussion of techniques is also included. Microwave chemistry has a more prominent place in synthesis today and now has its own section. More detailed discussion of HPLC methods was added. Instrumental techniques to evaluate enantiomeric composition are now covered. Discussion of several methods for the purification of solids has been added. Another topic that gets enhanced attention is the safe handling of pyrophoric chemicals. This includes some excellent Internet resources in the form of video demonstrations of crucial operations. Internet video has been tapped to demonstrate several other techniques as well. New appendices are provided that address solvent properties including freezing point, miscibility, and toxicity.

Software, Internet, and other electronic resources for synthetic chemistry are discussed wherever appropriate. Both SciFinder and Reaxys are now covered. Electronic laboratory notebooks are likely the future for many chemists and one currently available tool is summarized. Capabilities of mobile devices to do some pretty significant chemical informatics are described.

While I am happy to have increased the utility of the *Handbook* by these additions, I have also been concerned with keeping the overall presentation concise, so the information is most accessible.

The book has supplementary materials such as a reaction checklist, an Excel spreadsheet to predict flash chromatography separations, video links, and a solvents chart. These can be accessed online from the url http://booksite.elsevier.com/9780128095812/

Acknowledgments

I would like to thank several anonymous reviewers and all of my graduate students and postdocs, past and present, who commented upon the earlier book and the proposal for this revision. They made it far better than I ever could have on my own. Tom Morton and Dan Borchardt critically reviewed parts of the manuscript. I am grateful for *many* figures supplied by Ace Glass. I have appreciated working with all of the Elsevier professional staff, particularly senior editorial project manager Jill Cetel.

My professional career would not exist without the influence of my father, J.M. Pirrung, MD. He not only gave me chemical aptitude through his Alsatian genes, but also taught me the first rows of the periodic table (and to say perhydrocyclopentano-phenanthrene) before I was in kindergarten. He shared with me his work as an industrial chemist before turning to medicine and many other professional pursuits. I thank him for enabling the lifetime of gratifying work I have been able to do in chemistry.

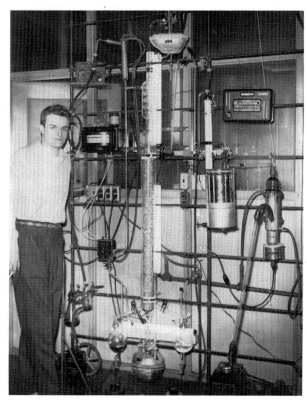

Source: From the personal collection of Michael C. Pirrung.

Above is a photo of my father at Kentucky Chemical c.1950 with a molecular still he built, his pride and joy. Some things in chemistry laboratories do not change (mantles, pumps, and dewars), even over quite a long time, but others clearly have—like the cigarette in his hand.

Safety

1

Chapter Outline

I strongly endorse the adage "safety first!," dictating that I begin this book with this important topic. Awareness and observation of all the best chemical safety practices are essential before undertaking any of the work described herein. The broadest general safety guidance concerning the hazards encountered in chemistry labs is provided by the text *Prudent Practices in the Laboratory* (Committee, 2011), which can be downloaded for free from the National Academy Press website. Texts like *Bretherick's Handbook* are also available that describe the hazards of a large number of specific compounds (Urben, 2007), and they too are available as electronic resources. Further, comments are made at numerous prominently marked places in this book about safety proscriptions for various compounds, processes, or equipment. These alone are hardly adequate preparation for entering the lab to perform synthetic work, however. Do not undertake any synthetic processes in a laboratory until you have been trained and certified in all aspects of safety that pertain to your work. Whether or not a specific alert is provided here concerning a particular topic, safety always must be foremost in the mind of the experimentalist. Finally, it is foolhardy and will likely put your health and life at risk to pursue anything described in this book in an "informal" laboratory setting like a kitchen or garage. In other words, do not try this at home.

1.1 Training

Essentially any organization in which the experiments discussed in this book will be performed will have several layers of formal safety training. All scientists should receive general institutional safety training. There will likely also be specialized training by department, if not subdiscipline. That is, some chemistry workers need to know everything about laser safety, but this is uncommon in the synthesis lab—likewise, most spectroscopists need not learn about peroxide-forming chemicals. Finally, the research group in which you work should have training that is specific to the types of chemistry that it performs. This training will certainly include written safety manuals and chemical hygiene plans. If your organization does not provide training at all of these levels, you should ask for it and do no experiments until you receive it.

Specific chemical safety training in the main hazard classes is also essential. For the synthetic laboratory, the chemical hazard classes typically include flammable liquids, acids, bases, peroxide-forming chemicals, strong oxidizing agents, strong reducing agents, water-reactive chemicals, pyrophoric chemicals, explosion risks, acutely toxic chemicals, and acutely toxic gases. Safety procedures regarding each of these classes are provided in sections of Appendix 1.

Emergency situations in the laboratory, fires, spills, or accidents, challenge all chemists to apply in the heat of the moment the training they have received, most typically from their environmental health and safety staff. Those staff, who are most practiced in chemical hygiene, are the best resources to provide this training, such as the use of fire extinguishers, safety showers, and eye washes.

Supervisors in the lab in which you work must identify the safety hazards present and provide a structure, in terms of physical measures, standard procedures, personal protective equipment (PPE), training, and laboratory rules, to minimize their potential effects on human health. The training you receive that is specific to your own lab will certainly include this information. Guides are available for the identification of hazards in chemistry labs (Hazards Identification and Evaluation Task Force, 2013), and are designed to be used by chemists at all levels of experience. Reviewing this material can provide a greater appreciation of all the matters that have been considered in developing the chemical hygiene plan for your lab.

1.2 Safety Data Sheets

For any hazardous (or potentially hazardous) substance in commerce, the vendor must make available a safety data sheet. This requirement is often met simply via an easily accessed online archive, enabling users to obtain safety information even before purchase. These sheets include components such as names/synonyms for the compound, its hazards, composition of the form provided, physicochemical properties, stability, handling and storage requirements, recommended exposure controls, toxicological data, ecological data, waste disposal, and first aid, firefighting, and accidental release measures. That being said, one safety data sheet found for sucrose (sugar) indicates that, in case of ingestion, 2–4 cups of milk or water should be given if the "victim" is conscious and alert. Of course, we are not aiming to chide vendors here—they are simply providing a statutorily mandated document that meets the vendor's obligation to inform users about the hazards of a product (fulfilling the "right-to-know" principle)—but one must use one's own scientific knowledge and judgment in interpreting the contents of a safety data sheet.

1.3 Safety Pictograms

To quickly and clearly communicate to everyone the chemical hazards that they may encounter, a variety of safety pictograms have been used over time and for various purposes, such as for transportation or emergency responders. These pictograms underwent a recent revision, and the set currently used worldwide is given in Fig. 1.1.

Figure 1.1 The nine safety pictograms in the globally harmonized system. (A) Harmful (includes skin/eye/respiratory tract irritation, narcotic effects); (B) Compressed gas (includes cryogens); (C) Health hazard (includes carcinogenicity, mutagenicity, reproductive/specific organ/aspiration toxicity); (D) Toxic (acute/severe); (E) Explosive (includes organic peroxides); (F) Flammable (includes pyrophorics, water reactives, organic peroxides); (G) Oxidizing; (H) Corrosive (includes attack on metal and skin, serious eye damage); (I) Environmental hazard.
Images from www.osha.gov.

Table 1.1 **Protection Provided by Various Glove Types**

Type	Do Not Use With	Resistant to
Latex	Organic solvents, amines	Acetonitrile, alcohols, aqueous
Nitrile	Aromatics, phenols, ketones	Solvents, oils, alcohols, ethers, some acids and bases, aqueous
Butyl rubber	Aromatics, hydrocarbons	Ketones, alcohols, phenols
Neoprene	Aromatics, hydrocarbons, amines	Acids, bases, alcohols, peroxides, phenols, acetonitrile
Viton	Ketones	Aromatic, chlorinated
Polyvinyl chloride	Ketones, amines, alcohols, phenols, ethers	Acids, bases, peroxides

1.4   Personal Protective Equipment

PPE is an important tool to minimize chemical exposure that affords other safety protections. Most synthesis labs have standards requiring a minimum level of eye protection (safety glasses with side shields) and greater protection (like goggles or a face shield) for more hazardous procedures. Full-length pants and fully enclosed shoes are also typical requirements. Lab coats appropriate to the tasks at hand are commonly mandated.

Wearing gloves is a choice often made by chemists even if not required by their lab's standard procedures. Gloves are available in a wide variety of materials and it is important to know the compounds that will be contacted to select the proper protection. Some of the main options are summarized in Table 1.1. There are also many other resources available to match an appropriate glove to hazards. Current information is available from vendor's websites, such as http://www.ansellpro.com, which includes compound-by-compound listings of the chemical resistance of different glove types. The commonly used latex gloves provide great dexterity but minimal protection, since they are barriers only to mild aqueous solutions. At the other extreme are silver shield gloves that resist most compounds. Nitrile gloves are often the default option that offer good dexterity when there is no specific reason to prefer a different type.

1.5   General

A hallmark of chemical laboratory safety is minimizing exposure to *all* chemicals, thereby minimizing the need for knowledge of the toxicity of any of them. Much of this protection is provided by the labs in which we work, particularly the fume hood. Of course, synthetic chemists are always creating new molecules whose toxicity has never been examined, so this is a double incentive to make sure we have minimal contact with them. As much of your work as possible should be performed in the hood.

Centuries ago, chemists smelled and tasted their products, but no more. Yet, there is no need for alarm about the potential health effects of new compounds; while there are examples of compounds synthesized for research that unexpectedly proved to be highly toxic and harmed the chemists who made them, this is extremely rare.

The training that begins in this book may someday lead a chemist using these skills to the manufacture of active pharmaceutical ingredients (APIs) for a drug. In settings where such compounds are prepared, their potent biological activity requires great care in handling, but the hazards of APIs are also highly characterized (after all, they will be administered to patients), so proper precautions should be well understood.

Chemists, especially synthetic chemists, can discover during the course of their research compounds or procedures that are hazardous, most typically by an accident in their labs. They often aim to prevent others from suffering the same fate by informing their community of the hazard in the most immediate way. When print media reigned, this meant writing a letter to the editor to be published in *Chemical and Engineering News* (C&EN), the weekly in-house publication of the American Chemical Society. These so-called safety letters established a permanent record of their findings. The editorial staff of C&EN recognized the value of these letters to posterity and created a Web page where all letters published since 1993 are archived—http://pubs.acs.org/cen/safety/. The compound classes found there include azides, oxidizing agents, nitro compounds, alkynes, and perchlorates, suggesting the types of chemicals of which chemists should be particularly wary. This is a good page to scan periodically in case you missed a safety letter originally. A Web resource on general chemical safety that is also curated by C&EN is available: http://cenblog.org/the-safety-zone/.

It is also a good idea to review some of the general experimental principles and techniques that were covered in the organic teaching laboratory before undertaking the more sophisticated and less predefined activities of the research lab. One of the best resources for this type of information is *The Organic Chem Lab Survival Manual* (Zubrick, 2016).

The foregoing admonitions notwithstanding, this book is *completely insufficient* to provide you with all of the guidance necessary to safely perform synthetic reactions in the laboratory. All other modalities mentioned, including reading relevant texts, hands-on demonstrations, online training, videos, and more, are essential. You should look to your supervisor for all relevant information about the hazards of the compounds, equipment, and procedures used in your lab.

References

Committee on Prudent Practices in the Laboratory: An Update, National Research Council, 2011. Prudent Practices in the Laboratory: Handling and Management of Chemical Hazards (Updated Version). National Academy Press, Washington, DC. http://www.nap.edu/download.php?record_id=12654.

Hazards Identification and Evaluation Task Force, 2013. Identifying and Evaluating Hazards in Research Laboratories. American Chemical Society, Washington, DC.

Urben, P.G., 2007. Bretherick's Handbook of Reactive Chemical Hazards seventh ed. vols. 1–2. Elsevier, Boston. http://app.knovel.com/hotlink/toc/id:kpBHRCHVE2/brethericks-handbook/brethericks-handbook.

Zubrick, J.W., 2016. The Organic Chem Lab Survival Manual: A Student's Guide to Techniques, tenth ed. Wiley, Hoboken.

Searching the Literature

2

Chapter Outline

Organic synthesis has the largest literature of any field of chemistry, making the searching of it a mammoth task, and developing a command of it a lifelong endeavor. When aiming to obtain a particular molecule, a good appreciation of how it has been prepared in the past is essential. Electronic data retrieval tools are ideal for finding this information, but the chemist should not totally ignore the classical print literature—not everything useful has made its way into electronic form. The simplest way to obtain a compound is to buy it, of course.

Chemical Abstracts maintains a database of specific molecular species called the Registry, and each entry gets a unique identifier called a CAS Registry Number, CASRN, or CAS number. That applies to stereoisomers including enantiomers, so there may be several entries for what most chemists would regard as one molecule, representing its different forms (*R*, *S*, racemic, of unknown stereochemistry, isotopically labeled forms, etc.). The CAS number is a universal molecular identifier that makes the search for a source of a compound more tractable, like the ISBN for a book. There are over 109 million compounds in *Chemical Abstracts* as of 2016, and of course only a small fraction of those are commercially available. Even among compounds identified as commercial, there can be significant differences in availability between those sold by major vendors in the United States and western Europe versus other parts of the world, where a great deal of excellent synthetic chemistry is being practiced.

2.1   Commercial Availability

The two most general resources used to locate commercially available compounds are SciFinder and Reaxys. The former is the electronic portal to the *Chemical Abstracts* database. The latter includes a major organic chemistry abstract series, *Handbuch der organischen Chemie* (colloquially known as *Beilstein*, after its founder). Many different types of searches using these resources are useful for the synthetic chemist; they make the

Handbook of Synthetic Organic Chemistry. http://dx.doi.org/10.1016/B978-0-12-809504-1.00002-9

information in the databases much more accessible than print, and it is unlikely that many current chemistry students have performed searches in the physical volumes. They also go far beyond the physical volumes, which did not identify commercial compounds. While a comprehensive description of the use of these tools is not intended here, a few points should be kept in mind. While both are incredibly powerful for modern literature searching, Reaxys likely has an advantage for older literature, pre-1965, since *Beilstein* has been published since 1881. SciFinder offers wider coverage since 1965, and more comprehensive coverage of organic reactions specifically since 1985. Because they are based on different databases and different search algorithms, the chemist should not be surprised that searches on the same compound on each portal will provide somewhat different results.

SciFinder is a browser-based tool that permits searching by chemical structure, either as an exact match, or based on substructure, or based on structural similarity. Structures can be drawn in SciFinder itself, or in another chemical drawing program and pasted into the structure window in SciFinder. Some chemical drawing programs also provide click through to a structure search in SciFinder. To avoid an unmanageable number of hits when searching by substructure, it may be necessary to limit the structure based on atoms that may be further substituted or rings that may be added. These parameters are set within SciFinder itself, with the {Lock atoms} and {Lock ring fusion or formation} commands (key icons in the tool bar of the structure editor) (Fig. 2.1). Searching for an exact match is tricky, as it is easy to make a trivial change

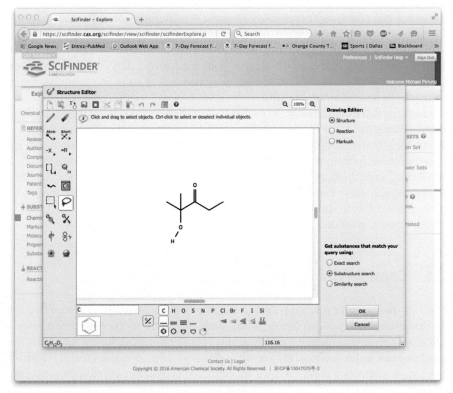

Figure 2.1 The structure search window in SciFinder.
Used with permission from CAS, a division of the American Chemical Society.

in structure that would seem obviously the same to a chemist, but is not regarded that way by SciFinder. For example, searching for a methyl ester would give no exact hits if only the ethyl ester is known. Better is a search by similarity. The hit list is ranked in decreasing order of similarity, so if there is a trivial difference between the query and a compound in the database, it will be one of the first hits. Seeing the known compounds that SciFinder regards as similar to the molecular query can also suggest better search strategies for the desired compound, or alternative compounds better suited to the synthesis goal that prompted the search. The hits in a structure search of any type can be limited to those that are commercially available through an option on the search page.

The substances identified in a SciFinder structure search are laid out in a grid, each cell headed by the CASRN (Fig. 2.2). In addition to the molecular formula and structure, up to three icons appear in the compound window. The first icon (of a folded page) is preceded by the number of literature references (publications, books, patents, etc.) that include the compound. The chemist should be skeptical of the reality of a molecule's existence if this is absent. It is a fair question how a molecule with no references could be entered into the database, but if so it likely exists only virtually, as a structure and not a compound. The second icon (of a flask) indicates that reactions involving the compound are found in the CAS database. This will be discussed further later. The third icon (of a flask with a price tag) shows that it is commercially available

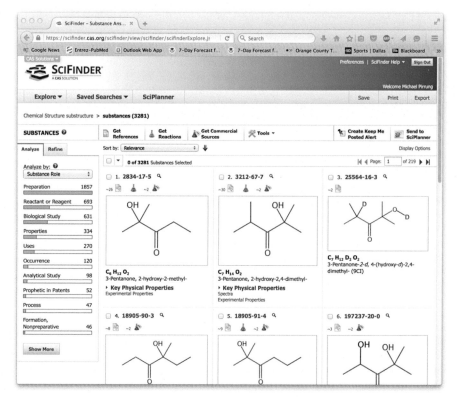

Figure 2.2 The results window from a structure search in SciFinder.
Used with permission from CAS, a division of the American Chemical Society.

and is preceded by the number of suppliers. Clicking this flask opens a window that lists the companies selling it and the quantities available. This last point is important. Quite a number of molecules have been registered with *Chemical Abstracts* by small companies who are aiming to sell compound libraries in tiny amounts for biological screening, or who are custom-synthesis firms who prepare a compound only when it is ordered. Their compounds, while meeting a strict definition of commercial availability, are not articles of commerce in the same sense as compounds stocked in bulk by major vendors. These include Sigma–Aldrich, Alfa Aesar, and Fluka, and specialty reagent houses like TCI, Maybridge, and Fluorochem. Information on less well-known suppliers (phone, address, URL) is available by clicking a link next to each supplier listing. Even if the catalog of a small supplier has been recently uploaded and SciFinder shows they keep a compound in stock (and even if the company's own website shows the compound is available), it is a good idea to contact the vendor directly to determine price and availability before concluding the compound can be purchased.

Reaxys has similarities to SciFinder in its user interface and capabilities. Given their nonidentical databases, searching both is advisable to ensure that all possible commercial sources of a compound are found. The results window from a search on the compound used in the SciFinder example above is shown in Fig. 2.3. The flask icon indicating commercial vendors is familiar. Click through from this icon may be to vendor's websites or to third-party databases of commercial compounds. The {Show Details} link gives access to a large amount of information on the compound, including physical properties, spectra, and references, with click through to publisher's websites.

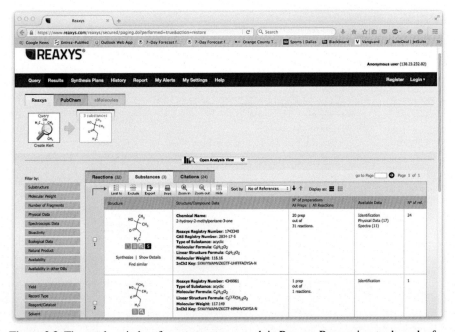

Figure 2.3 The results window from a structure search in Reaxys. Reaxys is a trademark of RELX Intellectual Properties, SA, used under license.

Reaxys and SciFinder are both subscription services, but there are also some free Internet-based commercial compound search engines. Internet sites can come and go at a rapid pace, so there is some risk that links to such resources will become outdated, but the aim to maintain a companion website to this book should keep them current. One resource worthy of mention is http://www.chemexper.com, which has a basic structural drawing window and returns a vendor listing in response to a search for commercial compounds.

It can also be worthwhile to search the websites of major chemical vendors for a desired compound. Sigma–Aldrich offers a resource to find its products by structure, among several options. A nice aspect of its search site is that it encompasses Aldrich's sister companies, including Sigma and Fluka. Aldrich is a vendor that has always provided excellent service to synthetic chemistry, and many chemists keep an Aldrich catalog at their desks because of the wealth of data it includes on all of the compounds listed. Much of this information, and some additional, is also on the website.

Structural entry on Web pages via drawing tools has become widespread but is not universal, in some cases due to incompatibilities between browsers and the plug-ins needed for particular tools (Java, etc.). If the drawing tool fails on a particular site, one alternative is to enter the structure as a SMILES string. Many approaches have been taken to manipulate chemical structures electronically, with one basic machine-readable nomenclature being the SMILES format. SMILES stands for simplified molecular input line entry specification. The SMILES description of a structure can be obtained from a drawn structure in a program like ChemDraw using the command {Copy as SMILES}. Upon pasting the SMILES string into the search page, the structure should appear.

2.2 Literature Preparations

If a compound cannot be purchased, there may be a known preparation of it. It is almost always preferable to at least begin with a method that others have used to prepare a compound, rather than trying to invent one from scratch or based on other compounds that may not be very similar. Identifying the specific publications in which a compound is a *product* among many more papers in which it has been mentioned in some other way is straightforward using SciFinder. The {Refine} command allows those sources to be selected. A nice feature of Reaxys enables rapid consideration of the best literature route to a target. It is accessed by clicking the {Synthesize} link for a molecule, which requires selecting a method; in this example, Autoplan was used. That search returns one or more literature routes to the target, which are viewed in separate tabs (Fig. 2.4). Also, for each preparation, a complete synthesis tree is shown in a small window, enabling the attractiveness of the whole route, rather than just the final reaction, to be quickly evaluated.

If a particular method to prepare a compound is sought, it may prove useful to search by reaction rather than by compound structure. This is done simply in SciFinder—multiple structures are entered into the chemical structure search window, and a reaction arrow is drawn from the starting material(s) to the product. SciFinder will then label the molecules it understands to be the reactant(s) and product(s) and searching for this reaction can be initiated. The number of hits is often quite large, but adding qualifiers

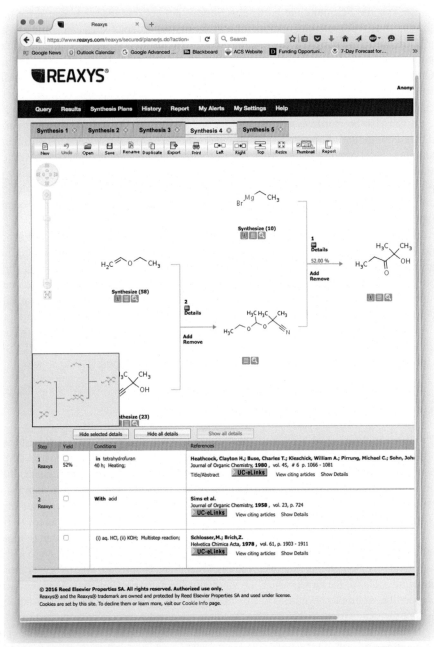

Figure 2.4 The result of a synthesis Autoplan in Reaxys. Reaxys is a trademark of RELX Intellectual Properties SA, used under license.

can reduce them to a manageable number. For example, reactions can be selected by applying the {Refine} command based on yield, or the number of steps, or the general type of chemical transformation. Another option that often reduces the possibilities is to identify which atoms of the reactant correspond to which atoms in the product. This may seem obvious to chemists, but it is not obvious to the software. A typical search may include results that do not involve the functional groups intended, but they can often be eliminated by this tactic. A nice feature of reaction searching in SciFinder is that brief summaries of experimental details are provided. This can be invaluable when the chemist cannot access the primary literature in which the full experimental description is found, which might be so because it is not available in electronic form or because it is not accessible at the chemist's institution. Likewise, Reaxys permits reaction searching based on an entry in the drawing window, with the ability to limit hits based on yield and reaction conditions.

Another way to search the literature for preparations uses citations to a compound's original report. Many institutions have access to the Web of Science, a database through which publications can be located that cite a particular reference. This is a good way to discover if other workers later made modifications or improvements in a procedure that would otherwise be difficult to locate. Alternatively, accessing the original publication on the publisher's website gives in some cases direct links to all subsequent papers that cite it. This is accomplished via a resource such as the Cited-by Linking service of Crossref, an organization of scholarly publishers that works to make content mutually easy to access. Of course, identifying a source that cites an earlier paper is no guarantee the newer literature includes the compound of interest.

2.3   Experimental Procedures

A preparation of a compound that was conducted in one's own laboratory is most preferred as a starting point for today's preparation. In some cases the chemist who actually did the prep may be available for consultation, or at least his or her research notebook pages can be consulted. A properly kept research notebook (see Chapter 8) is almost always more useful than any literature preparation because it gives details that never appear in a publication, such as actual chromatograms and spectra, drawings or pictures of apparatus, and properties of chromatographic or distillation fractions.

The best literature experimental descriptions will be found in the compilation *Organic Syntheses*. These procedures are generally aimed at synthesizing compounds on a fairly large scale (multigram or larger, rather than milligram). They are distinguished from essentially all other literature preparations by having been checked in the laboratory of another experienced synthetic chemistry group. The little details that can be important to success can be incorporated into the published procedure through the experience of the "checkers" in doing the experiment targeted for today, reproduce someone else's preparation. The only drawback of *Organic Syntheses* preps is that so few of the known compounds the chemist might need in research have been through its rigorous review process. Even if the exact target compound is not in *Organic Syntheses*, a prep of a related compound might provide a good starting point if today's target is not too different structurally. Structure, keyword, title, author, registry number, molecular formula, and chemical name searching of the compilation are available on the website for *Organic Syntheses* at http://www.orgsyn.org/.

Next preferred as a resource for the preparation of a desired compound is one that has been described in a "full paper" (which gives experimental details, as in *J. Org. Chem.*). These experimental details are increasingly being relegated to the electronic version of the journal, often called supporting material or supplementary information. The techniques for searching for this literature include *Chemical Abstracts* or *Reaxys* as described earlier, as well as specialized sites maintained by each publisher for its journals.

Least informative concerning the preparation of a target compound are so-called communication or letter publications, which do not include experimental descriptions. In some cases such details may be provided in Supplementary Information, but not typically at the same level as in a full paper (enumerated in Chapter 15). If there is no supplementary information, the authors still may have included some information about reaction conditions in the written narrative or on the reaction schemes. Also, it is always worth contacting the senior author to ask if experimental details could be provided. Electronic contact information is part of most modern publications, and authors of older literature may be readily located to make such a request. Still, the novice experimenter may have little to go on when working from these types of papers, and reproducing preparations from them is widely regarded as difficult at best.

In some cases, reactions or experimental procedures may be located in patents. However, compounds and procedures described there do not necessarily have a basis in reality. Patents can be prophetic, meaning no evidence is required that the inventors actually achieved what is described to receive a patent. It is only necessary that someone skilled in the art of organic synthesis would reasonably expect it could be achieved without undue experimentation. Obviously, the tentative nature of compounds and procedures in a patent is no basis for planning a reaction. That is not to say everything in patents is false, just that skepticism must be exercised.

Finally, it should be recalled that the electronic versions of journals are available most comprehensively for relatively recent editions. Electronic coverage of earlier compounds and reactions cannot be assured. Thus, traditional book-based literature search methods may be required if the reaction of interest has a long history.

2.4 Other Electronic Resources for Synthetic Chemistry

The Internet is dynamic and evolving rapidly, and also supplies some quite useful and incomparable resources. While there is a companion website for this book at which the links to referenced websites are expected to maintained, it may be useful to search for the resource directly if the link provided in print or on the site are not active.

A useful resource to locate information concerning specific reaction types is the *Encyclopedia of Reagents for Organic Synthesis* in its Internet form, e-EROS. The URL is http://onlinelibrary.wiley.com/book/10.1002/047084289X. It allows searching of a database of more than 50,000 reactions and more than 4500 of the most commonly used reagents by chemical structure and substructure, conditions, and reaction type.

Another significant resource is *Organic Reactions*, a text series that has been published since 1942. Its URL is http://onlinelibrary.wiley.com/book/10.1002/0471264180. Each chapter is essentially a review article focused on a single reaction type. All of the "name" reactions in organic chemistry (those important enough that they are referred to by the chemist(s) who originated and/or popularized them) and many more reaction types have been the subject of chapters in *Organic Reactions*. It is unique in requiring tables that collect all published examples of that reaction. Since some volumes in the series predate electronic literature, they may include references and examples that cannot be readily located by search engines, and for that reason alone they are an important adjunct to the electronic literature searching discussed in Section 2.2. Another appeal of these articles is that they are written by experts whose own research gives them unique knowledge and insight into that reaction, rather than relying on the algorithms and access to digital literature of search engines. *Organic Reactions* focuses on issues such as substrate scope, reaction limitations, stereochemistry, structure–reactivity relationships, and experimental conditions, and includes more than 200,000 individual reactions. These features make it an excellent resource for the chemist aiming to practice or troubleshoot a new example of one of these reactions.

There is a rapidly growing set of applications ("apps") related to synthetic chemistry for smartphones and other portable devices (both major operating systems) (Libman and Huang, 2013). They begin with ChemDoodle Mobile, which enables "digital" touch screen drawing (literally; that is, done with a digit) of chemical structures for entry into other chemistry apps. It also provides a nice range of calculable molecular parameters. The MObile REagents (MORE) app offers a large, searchable database of molecules and commercial products and includes the capability of direct optical structure recognition (OSR), side-stepping the need for structure drawing. OSR means taking a picture of a structure from a smartphone camera and automatically translating it into an electronic molecular representation. ChemSpider is a powerful structure-based search engine that provides properties, literature, and vendors for organic compounds, replicating in some aspects the traditional database search tools described Section 2.1.

Reference

Libman, D., Huang, L., 2013. Chemistry on the go: review of chemistry apps on smartphones. J. Chem. Educ. 90, 320–325. http://dx.doi.org/10.1021/ed300329e.

Reagents

3

Chapter Outline

It is wise to check the purity of all reactants before starting a reaction, especially when trying new reactions. Otherwise, if a reaction fails, it is difficult to know if the cause is the reagents or a fault in the procedure. Analytical methods could include titration, NMR spectroscopy, thin layer chromatography, or gas chromatography, depending on the types of contaminants that might be involved. Such baseline data on reactants will also be useful in analyzing reaction mixtures so that starting materials can be easily identified. Some reactive reagents (trimethylsilyl triflate, acid chlorides) may not be easily analyzed. The effectiveness of NMR spectroscopy for determination of reagent quality may be surprising. For example, the chemical shifts of the methyl groups of acetic acid and peracetic acid appear at δ 2.16 and 2.22, so an NMR spectrum of a commercially available peracetic acid solution readily provides its titer.

A classic text on the purification of reagents was long referred to around labs simply as Perrin, but the series was taken over a while ago by Perrin's coauthor Armarego. It is now in its seventh edition (Armarego and Chai, 2013) and available as an e-book, which may make it more accessible for some. A major part of this text is an amazing compound-by-compound listing of recommended methods for purification. Another resource is *Practical Organic Chemistry*, sometimes called simply Vogel (Furniss et al., 1989). Although this text is somewhat dated in terms of techniques and equipment, best practices for reagent purification change slowly. If a reagent should be purified, if it is not too reactive, and if it is somewhat volatile, distillation is a frequent choice.

Handbook of Synthetic Organic Chemistry. http://dx.doi.org/10.1016/B978-0-12-809504-1.00003-0

3.1   Short-Path Distillation

This method entails a simple distillation that is useful for quantities of material to be purified in the gram to several grams range. The short-path distillation apparatus is a simple, integrated still head that accepts a mercury immersion thermometer and receiver (Fig. 3.1), and can be placed under vacuum or an inert atmosphere. The pot is heated with an oil bath, and the apparatus is wrapped in aluminum foil to minimize radiative heat loss and the overheating otherwise necessary to get material to distill over. A crude fractionation of the distilland can be achieved by short-path distillation, provided the boiling points are well separated. To distill under reduced pressure, a multiple receiver adapter (Fig. 3.2; otherwise known as a "cow" for obvious reasons) can be used that permits the distillate to be directed, merely by rotating the cow, to different receivers. This apparatus can be adapted for fractional distillation of larger quantities of reagent by adding a column such as the Vigreux type (Fig. 3.3). The column must be thermally insulated. A typical height equivalent to a theoretical plate value for a Vigreux column is 10 cm. A 50-cm column is therefore capable of providing about five theoretical plates, which can acceptably separate (c. 95% pure) compounds with a boiling point difference of 30°C.

When performing any distillation at reduced pressure, it is important to have an estimate of the boiling point of the desired material, which of course is pressure dependent. It is possible to translate a known boiling point at one pressure to the boiling point at another pressure by using a vapor pressure nomograph (Fig. 3.4).

To use this nomograph, given the boiling point at atmospheric pressure, place a straightedge on the temperature in the central column. Rotating the straightedge about this temperature will afford the expected boiling point for any number of external pressures. Simply read the temperature and the corresponding pressure from the point the straightedge intersects the first and third columns. For example, choose a boiling

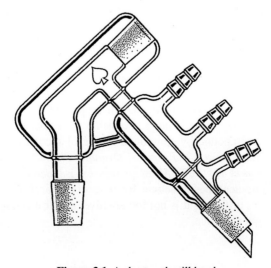

Figure 3.1 A short-path still head.

Figure 3.2 A cow adapter for vacuum distillation with collection of several fractions.

Figure 3.3 A Vigreux column for fractional distillation with moderate separating power.

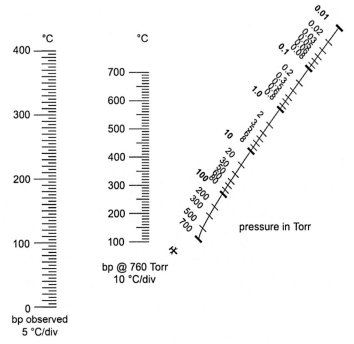

Figure 3.4 A vapor pressure nomograph permits the boiling point at any pressure to be predicted based on a known boiling point at a known pressure.

point at atmospheric pressure of 400°C. Using the nomograph and this temperature for reference, rotating the straightedge about this temperature will afford a continuous range of expected boiling points and the required external pressures necessary to achieve the desired boiling point. At a pressure of 6 Torr, the expected boiling point will be 200°C. Likewise, our compound boiling at 400°C at 1 atm would be expected to boil at 145°C at 0.1 Torr external pressure.

Electronic vapor pressure nomographs are available on the Internet, a nice one being: http://www.sigmaaldrich.com/chemistry/solvents/learning-center/nomograph.html.

3.2 Ampules

Some reagents (e.g., methyl triflate, ethylene oxide) are supplied in sealed glass ampules (Fig. 3.5). Two possible reasons for this packaging are that the compound is very reactive with the atmosphere, or the compound is quite volatile and conventional closures are inadequate to contain it. It is usually not a good idea to use only a portion of the material in the ampule and/or attempt to reseal it. The best strategy is to set up the reaction on a scale such that all of the reagents in the ampule are consumed in today's reaction. Second best would be to transfer the unused reagent into a sealable storage bottle or flask under an inert atmosphere. The neck must be broken off the ampule to

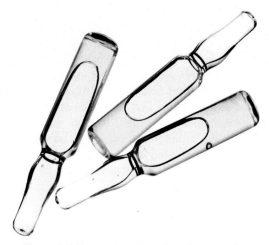

Figure 3.5 Reagents packaged in glass ampules.

access the reagent. Usually there is a marked line on which the break is intended to occur, but it is always a good practice to score the glass with a glass file or diamond glass cutter to be certain the neck breaks cleanly and easily. When the reagent is reactive with the atmosphere, this should be done with close access to an inert atmosphere or actually in an inert atmosphere (e.g., using the funnel technique of Chapter 5, Section 5.4, a glove bag, or a glove box). When the reagent is volatile, the bottom of the ampule should be cooled before opening to minimize the internal pressure.

3.3   Reagent Solutions

A wide variety of reagents are available from commercial suppliers as solutions. This makes for easy measurement of even quite air-sensitive reagents by volume using a syringe (see Chapter 9, Section 9.4). Reagent solutions are particularly useful for organometallic reagents that are better prepared on a large scale, but many, many air-sensitive and sophisticated reagents are today available in this way. One consideration against using these reagent solutions is their high expense on a molar basis (compared to preparing them in the lab), and another is the reliability of their titer. Even a new bottle from a supplier may have significantly different reagent concentration than the label claims. When a chemist accepts the titer of a purchased reagent solution, he or she is relying on the vendor to have gotten it right. The concentration listed is usually a minimum, but regardless, it is always a good idea to check the titer when possible. Of course, with storage and use over time, titer may decrease, so periodic titration may be necessary.

Bottles containing these compound solutions frequently come with closures intended to preserve an inert atmosphere, such as the Aldrich Sure/Seal. These involve a septum covered with a fiber mat fixed under a metal cap similar to that on a soda bottle (the crown cap). A conventional plastic screw cap goes on the bottle over this closure. The Sure/Seal works fairly well, but many chemists add their own precautions. The sleeve of a regular rubber septum (Chapter 5, Section 5.5) can be put on (upside down) over the Sure/Seal and copper wire used wire it on; a smaller second septum can be inserted inside the first and its sleeve folded over. Stretching of Parafilm M over this septum is also a good idea. Storage in a flammables-approved refrigerator helps to maintain concentrations of solutions in more volatile (i.e., pentane, ether) solvents. MeLi in ether is not stored in the refrigerator, as it precipitates out of solution and this changes the concentration. Most organometallics stay as clear solutions as long as they are good, with white precipitates (metal hydroxides) indicating some loss of reagent. All bottles of reagent solutions should be maintained with an inert atmosphere in the headspace (the region above the solution).

3.4 Titration

Organometallic reagents that are also strong, stoichiometric bases such as Grignard and lithium reagents may be titrated in several ways: against aqueous acid, against iodine, against diphenylacetic acid, and against alcohol. The first two work well with either type of reagent, whereas the latter typically work better with lithium reagents than Grignards, as Grignards are typically less reactive/basic than organolithiums.

The first method is performed by quenching an aliquot of the solution into water and titrating the hydroxide so produced with 1 M acid, using bromocresol purple or phenolphthalein as an indicator. This method has the disadvantage that it does not discriminate between the organometallic and hydroxide ion that may be present in the reagent solution from its reaction with moisture (Eq. 3.1). This hydroxide may be subtracted by using the Gilman double titration. Hydroxide can be measured independent from the organometallic because the latter is destroyed without the production of base by addition of dibromoethane (Eq. 3.2). The amount of hydroxide remaining is determined by titration as above. Another disadvantage of this method is the two-phase system that is produced when mixing organic and aqueous solvents. This makes the endpoint more difficult to determine than in a standard aqueous titration.

An alternative to the double titration involves reaction of a Grignard (or even an organozinc reagent) with iodine in a saturated solution of LiCl in THF (Eq. 3.3). This solvent gives a homogeneous solution (including soluble metal halide products) that enables the endpoint to be observed clearly (Krasovskiy and Knochel, 2006). This titration is self-indicating, as the brown color of the iodine is discharged at the equivalence point. The presence of hydroxide/alkoxides also does not interfere.

$$MeMgBr + H_2O \longrightarrow CH_4 + HOMgBr \tag{3.1}$$

$$MeMgBr + (BrCH_2)_2 \longrightarrow MeBr + C_2H_4 + MgBr_2 \tag{3.2}$$

$$\text{MeMgBr} + I_2 \xrightarrow[\text{0.5 M LiCl}]{\text{THF}} \text{MeI} + \text{IMgBr} \tag{3.3}$$

$$\tag{3.4}$$

Another method is performed as follows: recrystallized, oven-dried diphenylacetic acid is weighed and dissolved in anhydrous tetrahydrofuran in a round-bottom flask containing a stir bar and kept under nitrogen. The amount of diphenylacetic acid used is determined such that less than 1 mL of the organometallic solution (based on its estimated concentration) will be required, permitting a 1-mL syringe graduated in hundredths to be used for the titration. The syringe is filled (Chapter 9, Section 9.5) with the organometallic solution and it is slowly dropped into the flask. Once the diphenylacetic acid is fully converted to the carboxylate, further addition of organometallic produces the yellow dianion (Eq. 3.4), which indicates the endpoint. Because diphenylacetic acid can be neutralized by hydroxide in the organometallic, this process can lead to some error.

Another method uses an alcohol such as *sec*-butanol as the acid, and since it will protonate only the organometallic, no correction for hydroxide is necessary. The other key to this method is the indicator, phenanthroline or bipyridyl. These have the property that they form colored complexes with lithium and magnesium reagents. In their presence, a brown color is usually observed. In their absence, a yellow or clear solution is seen. The titration is performed with an automatic burette (Fig. 3.6) under an inert atmosphere. The automatic burette has a lower reservoir that can be filled with titrant (c. 0.6 M *sec*-butanol in anhydrous xylene, determined volumetrically). The section of the burette that we would recognize as a burette is filled with titrant by pumping a rubber bulb and forcing liquid up into the burette with pressure. It is a convenient apparatus to store and maintain the titrant solution and encourages frequent titration.

Anhydrous tetrahydrofuran (c. 3 mL) and the indicator (20 mg) are added to a round-bottom flask with a stir bar. A known amount of organometallic is added, and the color should develop. Titrant is added from the automatic burette until the color is just discharged. Provided that the endpoint is not greatly exceeded, replicate titrations can be performed in the same flask to obtain a more precise value.

A direct measurement of the concentration of a number of organometallics is available through proton NMR spectroscopy (Hoye et al., 2004). Solvent protons can simply be neglected because their signals differ significantly from the organometallic, and an internal standard of cyclooctadiene is used whose signals are readily discerned. The reagent solution is placed directly into an NMR tube containing a measured quantity of internal standard and the organometallic concentration is determined by comparative integration.

It is perhaps a concern that so many alternatives are available to titrate these reagents. In some instances, the availability of many solutions for a given task can indicate that no solution is optimal. Here, the numerous choices speak to the vast importance of organometallic reagents to organic synthesis.

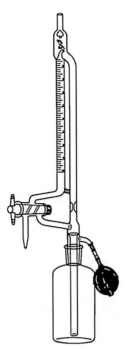

Figure 3.6 An automatic burette useful for the titration of organometallic reagents. The reservoir contains a nonhygroscopic solution of *sec*-butanol in xylene at a known concentration. The bulb is squeezed to fill the burette, and the tip is inserted through a serum cap into a flask under inert atmosphere and containing an ethereal solvent in which the titration is conducted.

It is also certainly possible to prepare solutions of reagents oneself, which will necessitate titration. A particular instance where this may be useful and for which commercial sources may not be available is volatile compounds. A classic example is anhydrous HCl in methanol. It is perfectly reasonable to prepare such a solution and keep it in the hood for several days, but the chemist should not store it for long periods because it is so corrosive. It is prepared by carefully bubbling gaseous HCl from a tank (see tank methods in Chapter 4) through a sparging tube into a known volume of dry methanol in an ice-cooled, tared flask. After the reaction has returned to ambient temperature, the flask is weighed and the HCl concentration is determined.

Safety Note

HCl has a substantial heat of solution, so bubbling it into a solvent is very exothermic. It is important to very carefully lower the sparging tube into the solution to avoid drawing methanol back into the gas line.

Other examples of useful reagent solutions include volatile organic compounds such as ethylene oxide or methoxyacetylene that have boiling points lower than about 30°C. It is very difficult to measure them neat in a syringe or on a balance, even when chilled, because they evaporate at such a rapid rate. Dissolving them in a reaction solvent and determining the titer, either by a direct method (e.g., integration in NMR spectroscopy) or by weight as was done for HCl/MeOH, can make their use much more convenient.

Sometimes repeated access (while maintaining an inert atmosphere) to homemade reagents such as Grignards that are not commercially available is required. One way to achieve this is with Mininert valves (Fig. 3.7) that are available in sizes that fit small (100 mL) reagent bottles, 5–30 mL reaction vials, and 14/20 joints. Another option is a three-way stopcock fitted with a joint and attached to a flask. To remove the reagent, a nitrogen line is attached to the sidearm of the stopcock. The stopcock is opened, a needle is inserted through it, and the liquid is withdrawn (see technique in a Chapter 9). The nitrogen flow maintains a blanket of inert atmosphere around the stopcock and replaces the volume of the withdrawn liquid.

Any time a reagent in any form is bottled or rebottled, an accurate label must be created that provides as many details as possible; it must include an accurate chemical name, concentration if a solution, hazard class(es), and date of bottling. This information is most secure and robust when written in pencil, since inks readily run when contacted by the pervasive organic solvents.

3.5 Reagent Storage

An extremely important principle of chemical hygiene that may be new to the novice lab worker is the segregation of chemicals that are from different hazard classes during storage. This applies to chemicals being stored before use, in a lab not being actively used, or after use in chemical waste. The aim of this requirement is to limit the range of hazards faced when a chemical in one hazard class is the source of an accident, and to prevent reaction with or release of chemicals from other hazard classes. The main hazard classes faced in the synthesis lab are acids, bases, strong oxidizing agents, strong reducing agents, explosion risks, flammable liquids, pyrophoric chemicals, water-reactive chemicals, acutely toxic chemicals, acutely toxic gases, and peroxide-forming chemicals. Some chemicals fall into multiple hazard classes and need to be segregated from both classes of which they are members.

It may be desirable in some settings (and is mandated in others) that chemical storage includes what is called secondary containment: a bin, bucket, beaker, can or tub to catch a compound should the primary container (flask, bottle, or can) somehow fail and release the chemical that is inside. The need for secondary containment applies particularly to storage in refrigerators, which can be seriously damaged by leaking chemicals.

For reagents that are obtained in standard screw-cap bottles, the cap should be wrapped in Parafilm M following first use. Parafilm is a thin, elastic sheet with low water permeability that is resistant to hydrochloric, sulfuric and nitric acids, potassium permanganate, sodium hydroxide, and ammonia solutions, and ethyl alcohol,

(A)   **(B)**

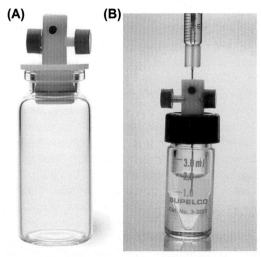

Figure 3.7 (A) A Mininert valve inserted into a vial. The small dot is a cylinder of rubber septum material through which a needle is inserted. (B) When the green button on the Mininert valve is pushed in, the pathway is clear for a needle to be inserted into the liquid to withdraw reagent. When the needle has been removed, the red button on the valve is pushed in to seal the container from the external atmosphere.

isopropyl alcohol, and acetonitrile. It is not stable to acetone or chlorinated hydro-carbons. The purpose of the Parafilm wrap is as much to keep the reagent in during storage as it is to keep the environment out. If a reagent is light-sensitive, the bottle should be wrapped in aluminum foil; brown glass bottles are rarely effective in fully preventing light penetration. If a reagent has a stench, the bottle should be placed inside a plastic bag.

Storage conditions depend on the particular reagent. Reagents that are not volatile, have low or no moisture or oxygen sensitivity, and are thermally stable can be stored on a shelf, ideally in a cabinet vented to the hood system. Reagents with moisture sen-sitivity but no oxygen sensitivity can be stored in a desiccator. Reagents that degrade when exposed to moisture or oxygen in the atmosphere should be stored under an inert gas. Reagents that are volatile or are otherwise heat sensitive should be stored in the refrigerator, which must be approved for flammables storage. Allow reagents removed from a refrigerator to warm to room temperature before opening the bottle, to avoid drawing air into the bottle. Few reagents truly require freezing, unless so labeled. Never place aqueous solutions in the freezer.

3.6 Subtle Reagent Variations

Some aspects of reagents can have a profound effect on their reactions, and in some cases these traits are uncontrollable. For example, many common organometallic reagents such as alkyl lithiums are purchased for convenience and because they can be hard to make. Does it matter whether the alkyl lithium was made by the commercial supplier from the bromide, the chloride, or the iodide? Does it matter whether the alkyl lithium is supplied in ether or tetrahydrofuran? Often, it does matter. Why should this be so? In some cases, it has been shown that dissolved metal salts (LiCl, LiBr, and LiI) affect the outcome of a reaction. These salts have different solubilities in ether and tetrahydrofuran (LiI is the most soluble), so the solvent (which will certainly be known) and alkyl halide precursor (which may not be known) of the purchased alkyl lithium can strongly affect the salts present in the solution and therefore the outcome of the reaction. The stereoselectivity of organometallic additions to carbonyl groups depends on such factors, for example. Wittig reactions can experience a similar effect, where the stereochemistry is dependent on the counterions of both the base used to generate the ylide and the phosphonium salt. Using sodium and potassium bases leads to salts that are insoluble and cannot affect the reaction. Such "salt-free" Wittig reaction conditions are commonly among the most stereoselective.

3.7 Dangerous Reagents

Certain classes of compounds are intrinsically unstable and/or prone to explosion. It is crucial to be aware of these classes, since examples might be prepared in the course of research that had never before been known. The chemist will not have a safety data sheet as a warning, but the explosion will be just as dangerous. The dangerous classes include acetylenes, acetylide salts, and polyacetylenes; hydrazoic acid and all azides, organic or inorganic (sodium azide is typically regarded as safe under reasonable treatment, but handling of it before or after the reaction can create hazardous materials such as hydrazoic acid); diazonium salts and diazo compounds; organic or inorganic perchlorates; nitrate esters of polyols; metal salts of nitrophenols; nitrogen trihalides; and peroxides.

Safety Note

CH_2Cl_2 (or any other haloalkane solvent) must *never* have any opportunity to come into contact with azide salts. Displacement of the two chlorines by azide ion will generate the dangerously explosive diazidomethane. Metal spatulas should not be used for transfers of solid azides, and organic azides should be protected from ground glass joints. Both can trigger explosions.

Azides are worthy of special mention. These compounds have recently been investigated much more avidly owing to the development of the copper-catalyzed azide-alkyne cycloaddition (CuAAC). This reaction (Eq. 3.5) is an early example of a research area now known as click chemistry, where bond formation occurs rapidly under ambient conditions. One method to prepare organic azides needed for click chemistry or other purposes is via substitution reactions using metal azide salts.

The explosion risk of organic azides has received serious consideration (Bräse et al., 2005). One indicator is the ratio of the number of carbon atoms to nitrogen atoms (three are in an azide group, of course). An organic azide with fewer carbons than nitrogens is an inherently dangerous explosive, and one with more carbons than nitrogens is still an explosion risk. Safer (to use the term loosely) organic azides would be those with at least three times the number of carbons as nitrogens (Bräse et al., 2005). Of course, these are simply rules of thumb; a chemist would not want to *lose* a thumb to an explosion of an organic azide that was supposed to be stable based on these guidelines. Therefore, the most judicious approach to the preparation and use of *all* azides is to apply all due safety precautions, including limiting the amounts handled and their concentrations, as well as the fanatical use of safety shields and personal protective equipment.

$$R^1{=\!=\!=}H + \overset{\ominus}{N}{\overset{\oplus}{N}}\underset{N}{\overset{}{N}}{-}R^2 \xrightarrow{Cu^{+1}} R^1 \overset{N=N}{\underset{N-R^2}{\diagdown}} \tag{3.5}$$

3.8   Reagent Properties

It is always useful to be aware of some of the physical properties and constants for the reagents used in a reaction. For example, an accurate value for the density can be very helpful in measuring reagents by syringe. Knowledge of the freezing points of all solvents and reagents is needed to avoid freezing them during low temperature reactions (freezing points of common solvents are provided in Appendix 4). This type of information is very often available. One source is the *CRC Handbook of Chemistry and Physics*, but other common resources are Reaxys, the websites of compound vendors, and chemical catalogs. These catalogs have incorporated more and more compound-specific information over time, and many chemists keep a chemical catalog or two at their desks. Aldrich even calls its catalog the *Aldrich Handbook*.

References

Armarego, W.L.F., Chai, L.L.C., 2013. Purification of Laboratory Chemicals, seventh ed. Butterworth-Heinemann, Waltham, MA.

Bräse, S., Gil, C., Knepper, K., Zimmermann, V., 2005. Organic azides: an exploding diversity of a unique class of compounds. Angew. Chem. Int. Ed. 44, 5188–5240. http://dx.doi.org/10.1002/anie.200400657.

Furniss, B.S., Hannaford, A.J., Smith, P.W.G., Tatchell, A.R., 1989. Vogel's Textbook of Practical Organic Chemistry, fifth ed. Prentice Hall, New York.

Hoye, T.R., Eklov, B.M., Voloshin, M., 2004. No-D NMR spectroscopy as a convenient method for titering organolithium (RLi), RMgX, and LDA solutions. Org. Lett. 6, 2567–2570. http://dx.doi.org/10.1021/ol049145r.

Krasovskiy, A., Knochel, P., 2006. Convenient titration method for organometallic zinc, magnesium, and lanthanide reagents. Synthesis 890–891. http://dx.doi.org/10.1055/s-2006-926345.

Gases

4

Chapter Outline

A variety of compressed gases are available in containers ranging in size from lecture bottles (Fig. 4.1; about the size of a 250-mL graduated cylinder) to small desktop cylinders (Fig. 4.2) to large tanks or cylinders (Fig. 4.3). Each type has a specific method for gas delivery.

4.1   Lecture Bottles/Small Cylinders

Lecture bottles come in several varieties, including traditional and modern versions, and differ for corrosive and noncorrosive gases. Traditional bottles usually have a female-threaded fitting into which a needle valve (Fig. 4.4) is directly attached. The valve must be stainless steel when used with corrosive gases. With the needle valve closed (fully clockwise), a nut below the fitting is screwed in (down, clockwise), which opens a pressure valve. Flow is then controlled with the needle valve.

The modern lecture bottle (Fig. 4.1) is more straightforward to use, with a tank valve to which a needle valve is attached. The needle valve is closed, the tank valve is opened, and flow is controlled with the needle valve as above. Regulators (see Section 4.2) are also available for use with lecture bottles. Their major drawback is that they are not returnable to the vendor when empty (with a couple of exceptions) and must be disposed of as hazardous waste.

The integrity of fittings is essential in assembling any gas delivery system; yet, this can be difficult to evaluate. A common plumbing technique, applying soapy water solution to all connections, readily detects gas leaks: soap bubbles form at fittings that are leaking. Each fitting should be tested with a leak-detecting solution once its section of the apparatus has been pressurized.

As mentioned in Chapter 3, Section 3.4 chemicals whose boiling points are near or below room temperature can be difficult to handle. Methyl chloride, propyne, trimethylamine, and vinyl bromide are examples of such compounds that are available in a liquefied state (under pressure) in small bench-top cylinders. These tanks contain from 275 mL to 2.2 L and are provided by Aldrich as Sure/Pak products. Because of

Handbook of Synthetic Organic Chemistry. http://dx.doi.org/10.1016/B978-0-12-809504-1.00004-2

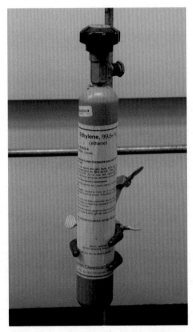

Figure 4.1 Lecture bottles used to hold and deliver small quantities of reagent gases.

Figure 4.2 Small tank in a gas delivery station used for propane and other liquefied gases.
© Sigma–Aldrich Co. LLC.

Figure 4.3 Large gas cylinders typically used for gases such as He, N_2, and H_2.
Photo provided by Airgas, Inc.

Figure 4.4 A brass needle valve for a lecture bottle.

their liquid storage form, the gas must evaporate to be delivered. The tank therefore becomes chilled upon gas delivery by the requirement for heat of vaporization. If significant quantities of gas are delivered, it is necessary to warm the tank to maintain the temperature, evaporation rate, and gas flow. This can be achieved with coils that circulate domestic hot water around the tank.

The cylinder fits into a station that includes a pressure regulator (Fig. 4.2), to which it is connected via a hose using a flare fitting. Like the Compressed Gas Association (CGA) fittings discussed later, these are designed to make a gas-tight seal through metal–metal contact. PTFE tape can be used to secure the threads of the fitting that hold the metal surfaces together but should not be applied at the gas connection itself. As is standard for pressure regulators, it is closed by turning its pressure-adjusting knob counterclockwise (fully out). The regulator outlet valve is also closed (fully clockwise) before the valve on the tank is opened. After opening the regulator (clockwise), the reaction vessel is connected via tubing to the outlet valve, through which gases are delivered at the pressure shown on the regulator gauge.

4.2 Tanks or Cylinders

Gas delivery from larger tanks or cylinders uses a specific fitting for each type of gas, with each fitting assigned a unique identifier by the CGA. The CGA numbers and the gases with which they are associated are listed in Table 4.1. Fittings differ in their size, the pitch of their threads, and even the direction they are threaded. This minimizes the chance an incorrect gas might be introduced into a system intended for another gas. It is particularly important there be no inadvertent mix-ups of reactive gases, and consequently their fittings are unique, whereas there is little difficulty using the same fitting for the inert gases. The chemist must be aware when multiple gases could be connected to a fitting on a gas system, and be certain that the contents of a tank that is being used to charge a gas system are correct.

Two types of gas delivery from large cylinders are available, the first more common for reagent gases. These are often in half-height tanks, which facilitates placing them into a fume hood. A simple needle valve screws onto the tank valve and has a hose adapter for output. The needle valve is closed (fully clockwise) and a hose is attached.

Table 4.1 **CGA Fittings for Compressed Gas Cylinders**

CGA#	Gases
320[a]	CO_2
330	HBr, HCl, H_2S
346	Air
350	B_2H_6, CH_4, CO, H_2
510	C_2H_2, C_3H_6
540	O_2
580	Ar, He, N_2
660	HF
705	NH_3

[a]Requires gasket.

Figure 4.5 A gas pressure regulator for a large cylinder.
Photo provided by Airgas, Inc.

The tank valve is opened by turning a square nut on the top of the tank counterclockwise with a crescent wrench. Delivery is controlled by the needle valve. Examples are NH_3 and HCl.

The second method is common for inert gases and chromatography gases, and these are typically provided in full-height tanks. Regulators (Fig. 4.5) are used to deliver gases from large cylinders at constant pressure. They adapt the tank pressure, which can be up to 3000 psig when full, to delivery pressures, ~100 psig. A regulator has two gauges and a knob or wing screw in front, and screws onto the tank valve. The regulator sometimes has a needle valve at the output, but if it does not, consider adding one. The cylinder valve and regulator have CGA fittings, in which a gas-tight seal is made by metal–metal contacts inside the fitting itself, not through the threads. Therefore, *no* sealant, pipe dope, or Teflon tape is used anywhere on these fittings, as it can enter and damage the regulator. If the fitting leaks, this will not be mitigated by thread tape; it must be repaired. In some cases, a plastic gasket is used between the two metal surfaces inside the fitting. After attachment of the regulator to the cylinder valve, the knob or wing screw is turned to the counterclockwise, fully closed position. The tank valve is opened, causing the right-hand gauge to show the pressure of gas contained in the tank. With the needle valve closed, screwing the knob or wing screw in (clockwise) depresses a diaphragm that opens to release the gas. The left-hand gauge shows the delivery pressure, adjusted by the regulator knob or wing screw. Opening the needle valve delivers the gas. Examples are H_2, He, and air for gas chromatography; N_2 and Ar for inert gases and HPLC degassing.

4.3 Gas Safety

General safety precautions must be observed whenever the chemist is working with gas tanks or cylinders. They must be held secure at all times unless they are being transported. This admonition applies to cylinders from the smallest to the largest.

Large cylinders should be chained or strapped to a wall or table at two points, and smaller cylinders should be clamped to the equipment rack in the fume hood or their storage location. Large tanks must be transported only with a tank cart. Any tank not in active use should bear its protective, screw-on metal cap. Tanks must never be moved, not even a meter or less, without this cap. Toxic and corrosive gases (e.g., NH_3, HCl) are best obtained in small tanks that can be placed in hoods. Any gas at high pressure carries the risk of asphyxiation, since if all of the gas in a tank were to be released in a closed area, its volume would be large enough to displace all the air.

Safety Note

Since in most cases labs rent their large compressed gas cylinders, which can be owned by the institution or a gas supplier, their proper maintenance may be the responsibility of individuals outside the chemist's immediate lab. Any hazards noted with the cylinders themselves should be reported promptly to the institution. Commercial suppliers also maintain emergency response systems that can assist workers in dealing with leaks or other failures of cylinders. It is always a good idea to know who your gas supplier is if you observe any difficulty with gas tanks.

Reactions on a Small Scale—1–25 mmol

Chapter Outline

Given the title of this chapter, something should be said about how the chemist selects the scale on which a reaction is conducted. One principle is that a reaction should be conducted on as small a scale as possible. This is so that, should the reaction fail, the loss of starting materials could not be too costly. "Cost" in this context can be literally the dollars spent for the purchase of starting materials or, more seriously, the chemist's time and effort in making the starting materials if they are the product of other synthetic reactions. For the latter reason chemists who are executing lengthy syntheses become highly adept at conducting reactions on small scales (a milligram or less!) "at the frontier." Of course, once the chemist has success with a particular reaction, the issue of scaling up to bring more material through the synthesis arises. An oft-mentioned rule of thumb is that a reaction should never be scaled up more than 4 times because it may not work as well at the larger scale, and further modifications of the procedure may be needed. This rule is likely honored as often in the breach as the observance, since it is obviously difficult to accumulate a significant quantity of a target compound if scaling up starts at the 1 mg level and is limited to fourfold increases.

Why should the outcome of a reaction be dependent on the scale at which it is conducted? There are several reasons. Mixing and heat transfer may be less efficient when reaction volumes are larger. Heterogeneous reactions that depend on reagent transfer between two phases are notorious for problems in scale-up. While a chemical step might have been limiting at the smaller scale, the efficiency of mixing might become limiting at the larger scale. The apparatus used for a reaction can never be identical at different scales—think of scaling up a 10-mL syringe by 100 times. For the same reason reagent concentrations may be increased upon scale-up to obtain greater productivity from a reaction, which may affect the yield.

Handbook of Synthetic Organic Chemistry. http://dx.doi.org/10.1016/B978-0-12-809504-1.00005-4

While economy suggests conducting a reaction on a small scale, some techniques for performing reactions and purifying products, especially those including processes with intrinsic material transfer losses, such as distillation, work far better on a larger scale. The choice of scale may therefore be limited by the need to use such a method in the reaction at hand. When working from a well-described literature procedure, it may be desirable (subject to the foregoing caveats) to try to follow its scale closely.

5.1   Reaction Flasks

Many reactions are run under an inert atmosphere of nitrogen or argon, are operated at other than room temperature, and require addition of reagents without exposing the reaction to the atmosphere. These three functions are well served by a three-neck flask (Fig. 5.1). The flask may be connected to an inert atmosphere manifold (see Section 5.4) through a gas inlet, fitted with a rubber septum for addition of reagents by syringe, and fitted with a thermometer that extends into the reaction solvent. This latter point is essential not only to observe the *exact* temperature of the actual reaction solution (rather than just the bath that surrounds the flask), but also to observe changes in reaction temperature as it proceeds. That is, upon the addition of a reagent, an exothermic reaction will show a temperature increase that can be observed if there is a

Figure 5.1 A glassware setup for a typical reaction.

thermometer in the solution. Because of the large thermal mass of a temperature-controlling bath, such exotherms are not easily observed via the bath temperature.

It is important to consider final reaction concentration, scale, and therefore volume in selecting flask size. Reasonable concentrations for fast bimolecular reaction rates might be in the 0.1–1 M range, so if a 10-mmol reaction is being set up, the final volume will be from 10 to 100 mL. On the other hand, reactions in which processes with unimolecular kinetics are to be favored over those with bimolecular kinetics (e.g., an intramolecular vs an intermolecular reaction) might be conducted at high dilution, meaning final concentrations closer to 0.001 M. To prevent sloshing of contents during mixing and to potentially allow the workup to be conducted in the reaction flask itself, the flask should be at least double the volume of the reaction, and ideally treble its volume.

Almost all reactions will be conducted in a glass apparatus. The most often used glass is Pyrex, a borosilicate glass that is one of the most inert substances known to chemical reagents. The only exceptions to this generalization are hydrofluoric acid (HF), phosphoric acid, and hot alkali. Of these, HF is the most serious problem. Attack on the glass can occur even when a solution contains only a few parts per million of HF. Other fluoride reagents can attack glass but usually to a lesser extent than HF. These reagents dissolve glass because of the very strong Si—F bonds that are formed. Phosphoric acid also corrodes glass. Alkali solutions up to 30% by weight can be handled in glass safely at ambient temperature but not at elevated temperatures. For reactions that use reagents that attack glass, apparatus made of Teflon or other plastics can be used. Borosilicate glass can be used safely at temperatures up to 232°C provided it is not subjected to rapid changes in temperature (>120°C). The strength of borosilicate glass increases with decreasing temperature, and it can be safely used at cryogenic temperatures.

A high-performance alternative to Pyrex is Vycor, a 96% silica glass. Many familiar pieces of laboratory glassware made from this material are available. Vycor has much more robust thermal properties than Pyrex, with an upper working temperature of 900°C, similar to fused quartz. Another key difference between Vycor and Pyrex is their absorption of UV light. A 1-mm thick pane of Vycor has 60% transmittance at 240 nm, whereas Pyrex has 60% transmittance of light only at 290 nm. This feature can be important in photochemical processes.

Safety Note

Never use glassware that is etched, cracked, star-cracked, chipped, nicked, or scratched, which makes it more prone to breakage, especially under vacuum, pressure, heating, or cooling. A master glassblower may be able to repair glassware that is damaged in these ways and can make recommendations concerning its safe use.

While the foregoing considerations address whether the glass can tolerate the reaction, the complementary consideration is whether the reaction can tolerate the glass. A number of reactions are dependent on the exact properties of the flask surface. The reasons for these observations are rarely established, but they are usually thought to be due to trace levels of acid on the glass from previous reactions or cleaning protocols. The chemical structure of glass includes surface Si—OH groups, but the acidity of Si—OH groups (in silica) is fairly weak (pKa of about 7, over 100-fold less acidic than acetic acid). It is unlikely the glass itself can provide acidity that affects many chemical reactions.

Two common methods are used to try to prevent interference with a reaction by the glass: base washing and silanization. Base washing is quite simple: typically, 28% aqueous ammonia solution is used to wash out the flask just before drying it directly in the oven (i.e., without a rinse). Silanization converts the Si—OH groups to Si—O—SiR$_3$ groups. The reagents used for glass silanization include dichlorodimethylsilane, trimethyldimethylaminosilane, and chlorotrimethylsilane. The former gives a more robust protection of the silicon oxide surface because it is attached by two Si—O—Si bonds.

Silanization is performed with a 5% (v/v) dichlorodimethylsilane solution in toluene or dichloromethane made up on a daily basis. Glassware at room temperature can be soaked in this solution, or it can be filled with this solution. After a 5 min to 2 h treatment, the glassware is removed from the solution, or vice versa. The glass is immediately rinsed with pure solvent. Exposure to moisture in the air is minimized by proceeding immediately to the next step. The glassware is covered or filled with methanol for 15–30 min. The methanol is drained off and the glass is blown dry using nitrogen.

5.2   Stirring

Certainly the most common (and easiest) stirring method is with a magnetic stirring bar and motor or stirring plate. This method works well for reactions in free-flowing solutions. Magnetic stirring bars come in a wide range of sizes, shapes, and coatings. Cylindrical stir bars with a fulcrum in the middle are excellent for Erlenmeyer flasks, but they do not spin well in round-bottomed flasks. For these containers, football-shaped stir bars (available for flasks 25 mL and larger) often perform better. Shorter cylindrical stir bars sometimes fit in round-bottomed flasks and spin well. For 10 mL or smaller flasks, vials or test tubes, tiny cylindrical stir bars (often called "fleas") work well. The most common stir bars are Teflon coated (Fig. 5.2). Teflon is a very tough substance, and these stir bars will stand up to almost any reaction conditions, except very strongly reducing conditions (i.e., alkali metals in ammonia, LiAlH$_4$). These reagents will rapidly turn Teflon-coated stir bars black. They still seem to work all right, so workers who do these types of reactions often keep their black stir bars just for this type of reaction. Alternatively, glass-encapsulated stir bars (Fig. 5.3) can be used.

Magnetic stirring may not work well, typically where the forces of friction, inertia, and solution viscosity are greater than the force exerted on the stir bar by the magnet in the magnetic stirrer. When this occurs, strategies to solve the problem include moving the stirrer closer to the bottom of the flask, using a stirrer with a stronger magnet, or using a larger stir bar. The magnetic stirrers with strong and weak magnets are often well known to all the chemists in the lab. Magnetic stirring may fail completely for reaction mixtures that are viscous, in which precipitation occurs, or that have larger reaction volumes. In such cases mechanical stirring may be required. A glass rod with paddles on its end is inserted through a bearing (oiled with mineral oil) in a ground-glass joint (Fig. 5.4). A variable-speed electric motor clamped above the flask is attached to the glass rod via a short piece of rubber tubing (slipped over it)

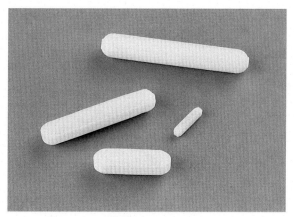

Figure 5.2 Teflon-coated magnetic stir bars.

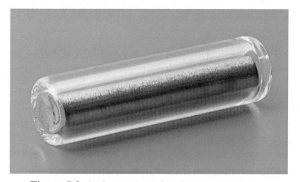

Figure 5.3 A glass-encapsulated magnetic stir bar.

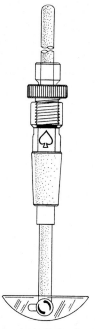

Figure 5.4 An overhead mechanical stirring bearing for a ground-glass joint.

Figure 5.5 A Variac or Powerstat sets the output voltage to AC heating devices to control their temperature and motors to control their speed.

or through a flexible shaft. This provides some flexibility in the connection so that the vibrations and movement that are natural to the operation of the motor do not place stress on any of the glass in the apparatus. The speed is controlled by the motor's integral control or a Variac or Powerstat (Fig. 5.5). A Variac is a variable AC voltage transformer (like a heavy-duty laboratory light dimmer).

Finally, it should be kept in mind that stirring affects only the rate of mixing of liquids and cannot affect the intrinsic rate of the reaction. Therefore, for reactions in a homogeneous solution, stirring may be unnecessary. It is rare that an experimental description omits stirring, however.

5.3 Glass Joints

Several considerations apply to use of ground-glass joints. In the past it was essential to use a good stopcock grease on every joint to ensure that the apparatus could be disassembled following the reaction, especially one involving a temperature change. These greases are typically silicon polymers that are quite soluble in hydrocarbon and halocarbon solvents, so it is easy to contaminate a reaction product with stopcock grease. This may be indicated by a yield greater than 100%, a high R_F spot on TLC, or indicative signals in crude NMR spectra (such as tetramethylsilane, the Me—Si signals from the grease are around 0 ppm (δ) in the proton NMR spectrum). An effective (though expensive) alternative is Teflon sleeves (specific to each joint size) or Teflon tape. For some applications, glycerin can be used as a lubricant and can be washed off with water after the reaction. Some glassware may be supplied with small glass "dogs" or hooks adjacent to each joint that enable the apparatus to be firmly held together with rubber bands, springs (sometimes supplied), or even copper wire. Alternatively, clips are available (Fig. 5.6) to hold the joint together, though these are likely less reliable and robust than the dogs and can sometimes interfere with other glassware and joints on the apparatus.

Figure 5.6 Metal clip for a ground-glass joint.
Photo courtesy of Kimble/Kontes.

5.4 Inert Atmosphere

Reactions are potentially susceptible to three minor components of air: oxygen, water, and carbon dioxide. Organometallic reagents are particularly susceptible to reaction with oxygen (Eq. 5.1) and water (Eq. 5.2). Any reaction involving a basic reagent more basic than hydroxide ion may be impeded by the presence of water (Eq. 5.3). Thus knowing the relative pKa's of a variety of organic functional groups and reagents is important. The acidities of a wide range of functional group classes in both aqueous and nonaqueous media are provided in Appendices 8 and 9. Carbon dioxide reacts with hydroxide (Eq. 5.4), alkoxides (Eq. 5.5), and amines (Eq. 5.6). Because they are gases, keeping oxygen and carbon dioxide away from a reaction is mainly a matter of using an inert atmosphere, as described later. Water may be present in microscopic amounts even on apparently dry glassware and other apparatus. Water is typically removed in one of two ways. Equipment can be dried in a drying oven (>130°C) overnight and then cooled in a desiccator. Some workers take glassware directly from the oven to the reaction bench (using insulated gloves or tongs) and cool it under a flow of inert gas. For equipment that was not dried in a drying oven, it may be dried with a flame (or, in labs that discourage open flames, by using a heat gun). Typically this method is used for a completely assembled apparatus, and offers the virtue that its components that are not usually oven dried, such as Teflon tape and serum stoppers, become dried. To be absolutely certain there is no significant water in an apparatus, it can be cooled below −40°C. Any trace of condensation signifies that water is present and the drying process must be repeated.

$$PhLi + O_2 \longrightarrow Ph\diagup{}^{O}\diagdown_{O}\diagup Li \tag{5.1}$$

$$PhLi + H_2O \longrightarrow PhH + LiOH \tag{5.2}$$

$$t\text{-BuOK} + H_2O \longrightarrow t\text{-BuOH} + KOH \tag{5.3}$$

$$KOH + CO_2 \longrightarrow KHCO_3 \tag{5.4}$$

KOMe + CO$_2$ $\longrightarrow$ MeOCO$_2$K   (5.5)

2 MeNH$_2$ + CO$_2$ $\longrightarrow$ MeHN−CO$_2^\ominus$ $\oplus$ H$_3$N−Me   (5.6)

The selection of nitrogen or argon as the inert atmosphere is based on a few considerations. Argon is about 3 times more expensive than nitrogen, and argon can only be obtained from a tank, whereas nitrogen can come from a tank or "house" nitrogen, usually obtained by the evaporation ("boil-off") from a liquid nitrogen dewar. Argon is denser than air, which is advantageous in that argon sinks to the bottom of any container and displaces the air. Argon is also not reactive with Li°, whereas nitrogen does undergo a slow reaction with Li°. Nitrogen is usually used without problem for Li°/NH$_3$ reductions, but the formation of organolithiums from alkyl halides and Li° is properly done under argon.

Another situation where argon use is preferred is a technique to protect a reagent that is not especially reactive with the atmosphere so it can be weighed or transferred without using more stringent and laborious precautions. This process simply inverts a glass funnel, held in a clamp, over the container. The funnel is connected with tubing to a source of inert gas to provide a bath around the equipment. The heavier-than-air argon is more effective than nitrogen at displacing the ambient atmosphere.

Labs committed to *extensive* work in an inert atmosphere, such as organometallic chemists, routinely use glove boxes, but they will not be covered here. Less effective and less convenient but far less costly substitutes are glove bags such as the Atmosbag (Fig. 5.7). This is an inflatable polyethylene chamber with integrated gloves and

Figure 5.7 This mid-sized glove bag is 39×48 in and has a 22 in diameter opening to load equipment.
© Sigma–Aldrich Co. LLC.

a gas-tight zipper to bring materials into and out of the bag. Gas ports and electrical pass-throughs are also provided. Glove bags are useful for the transfer of water-reactive solids into a flask, for example, which is otherwise difficult. A top-loading balance also can be put into the glove bag for use in an inert atmosphere.

Manifolds are very helpful to running reactions under inert atmosphere and can be quite simple to set up. They require a source of pure gas (nitrogen or argon). A very simple inert gas manifold is assembled starting with a gas drying tower (Fig. 5.8) filled with DRIERITE, a calcium sulfate desiccant that can be obtained with a color self-indicator, or KOH (never with bubbling through H_2SO_4). It is connected via tubing to a series of three-way stopcocks and terminates with an oil bubbler (Fig. 5.9). Tubing emanates from each stopcock, terminating with syringe needles or gas inlets (Fig. 5.10). The bubbler permits the chemist to see that there

---

Safety Note

It is crucial that reactions not be conducted in closed systems without using extreme precautions. The concept of not heating a closed system, as a practical illustration of the gas laws ($PV = nRT$), is impressed on chemistry students at an early stage. However, even without heating, conducting a reaction in a closed system can be hazardous, specifically any reaction that does (or could) produce a gaseous product or by-product. Some common examples include decarboxylation reactions and those involving highly nitrogenous molecules such as azides (RN_3) and diazo compounds ($R_2C{=}N_2$). Application of the gas laws makes this problem readily understood. Recall that a mole of a gas occupies the relatively large volume of 22.4 L. Any reaction that produces a gaseous product in a closed system must substantially increase the pressure because the volume is fixed. Such pressure increases can easily be large enough to separate ground-glass joints, even those that are held together with rubber bands, springs, clamps, or even wire. The possibility of spilling reaction mixtures that have their own intrinsic hazards is reason enough not to conduct reactions in closed systems. In a worst-case scenario, the pressure jump may be sufficient to explode the apparatus. This can occur even without flammable reactants, but the consequence is no different—shards of glass and reaction mixture shot around the lab. To attempt to address the possibility of reaction explosion for any cause, safety shields made from a variety of tough, clear plastics are available (Fig. 5.11). These are no substitute for a carefully planned and executed reaction, however.

---

is positive pressure and a moderate flow of gas in the system and permits volume/pressure increases to be quickly equilibrated. A bubbler provides a means to have a system isolated from the atmosphere that is not a "closed system."

An integrated glass manifold combines these elements in one apparatus. A more sophisticated manifold has dual chambers, one of which is attached to the inert gas

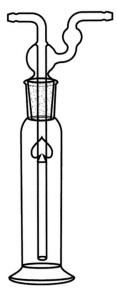

Figure 5.8 A gas drying tower is filled with a porous desiccant.

Figure 5.9 An oil bubbler filled with mineral oil shows that there is a positive pressure of an inert gas in a system and how fast gas is flowing through the system.

Figure 5.10 A gas inlet that can be used to attach a flask to the inert atmosphere manifold.

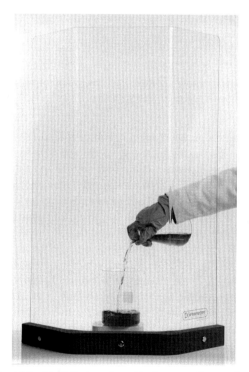

Figure 5.11 A plastic safety shield that can be set up between the reaction and the chemist. Photo provided by Bel-Art - SP Scienceware.

(A)

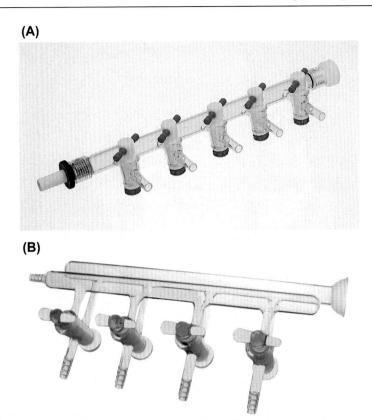

(B)

Figure 5.12 (A) A simple manifold to deliver inert gas to several vessels. (B) A double manifold that can deliver inert gas and apply vacuum.

source and the other is attached to a vacuum source (Fig. 5.12). In this setup a double oblique stopcock enables the lines running to the apparatus to be connected to one manifold or the other. This is a sure method of filling the apparatus with an inert atmosphere: evacuation and then filling. When using this technique, it is essential to have a fairly rapid gas flow into the nitrogen side of the manifold, as evidenced by vigorous bubbling, and ideally a large manifold volume, to avoid sucking air into the manifold when it is exposed (gradually!) to the evacuated flask by rotating the stopcock from one manifold to the other. Of course, when applying vacuum via flexible tubing/hoses, it is essential to use a tubing with rigid walls that will not collapse under reduced pressure. That is, conventional Tygon tubing will not serve the purpose.

Another feature of Tygon tubing that should be kept in mind is that its flexibility is maintained by a "plasticizer," usually a diester derivative of phthalic acid (e.g., dibutyl phthalate, dioctyl phthalate). These esters are quite soluble in many organic solvents, and in particular chlorinated solvents such as dichloromethane. Therefore, Tygon tubing must never be used to transfer solvents, and care must be taken when using Tygon tubing to connect gas sources to vessels containing such solvents because the plasticizer will be leached out of the tubing. The two consequences are that the tubing will

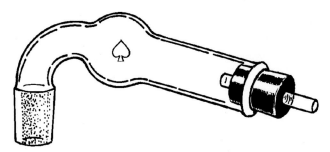

Figure 5.13 A drying tube that can be filled with desiccants such as CaCl₂ or molecular sieves.

become somewhat stiff and, more importantly, that the plasticizer will end up in the reaction mixture and contaminate the product.

Some reactions must be protected from atmospheric moisture but should not be attached to the inert gas manifold. For example, a reaction such as a Friedel–Crafts acylation produces HCl, which will likely escape the solution as a gas. HCl should not go into the nitrogen lines. For such situations a classical drying tube may be more appropriate (Fig. 5.13). Desiccants used in a drying tube might include the aforementioned DRIERITE, calcium oxide, barium oxide, or molecular sieves. Plugs of glass wool are used to hold the desiccant in the tube while permitting gas flow. Even when using a drying tube, the apparatus could be filled with dry nitrogen before reaction is begun.

Finally, an alternative to using a manifold to supply inert gas is a balloon (standard latex party balloons work fine, pulled onto a rubber stopper or a short length of rubber tubing). The balloon is generally connected to the apparatus through a gas adapter or even sometimes a needle (despite the obvious incompatibility). This approach is not as effective as a manifold because the volume of gas that can be delivered is much less. Balloons may also be used to provide "makeup" gas to a reagent bottle that makes up for the volume of the liquid that is being withdrawn.

5.5   Apparatus for Addition

Access to a reaction underway can be provided by stoppers or rubber septa. Stoppers can be glass or Teflon. Apparatus including Teflon stoppers should not be suddenly exposed to heat, since Teflon expands thermally much faster than glass. Conversely, stuck Teflon stoppers can be freed by immersing them in dry ice/acetone. One drawback of stoppers is that if the reaction experiences a sudden pressure increase, they can become projectiles. More often flexible rubber septa (also called serum caps or serum stoppers) are used for addition of reagents under an inert atmosphere (Fig. 5.14). They fit the internal diameter of the ground-glass joint snugly and have a flexible sleeve that is pulled down over the outside of the joint. This friction seal is fine to contain gases but still does not prevent pressure pulses from firing the serum stopper off the flask. They can be secured with copper wire. Reagents can be added to the flask as a solution via a syringe and needle through the serum stopper. Serum stoppers cannot tolerate a

Figure 5.14 Rubber septa for some of the most common sizes of ground-glass joints. The narrow, hard end is inserted into the joint and the wide, more flexible end is folded down over the outside of the joint.
© Sigma–Aldrich Co. LLC.

great deal of heat, and are not usually dried before use. As organic polymers they are also not very hydrophilic, so the amount of water introduced into reactions by serum stoppers is relatively small, and can only affect small-scale reactions. Serum stoppers can be dramatically swollen by organic solvents such as tetrahydrofuran, toluene, or dichloromethane. Care should be taken in using serum stoppers when they will be exposed to such solvents for long periods, especially at reflux.

Facilitating the addition of reagents to a reaction while maintaining an inert atmosphere can be done with several devices, or not. Provided that good inert gas flow is maintained, that the reaction itself is not too sensitive to the atmosphere, and that the reagent can be added all at once, it may be possible to simply remove a stopper, add the reagent (possibly via a simple glass funnel), and quickly replace the stopper.

The addition on a small scale of a viscous reactant that cannot be measured by volume to a reaction under an inert atmosphere creates a particular challenge. The best approach is to prepare and transfer a solution of the compound in the reaction solvent. This can be done by evaporating a solution of the compound into a tared 5–10 mL pear-shaped flask, the type that comes to a sharp point (maximizing the volume of solution that can be drawn into a syringe needle). The flask is fitted with a serum stopper and placed under an inert atmosphere. A minimum volume of a dry solvent is added (typically by washing it down the walls of the flask) and the compound is dissolved by swirling the flask. Another syringe is used to transfer the solution into the reaction flask. Half of the original volume of solvent is added to the pear-shaped flask and the residue is dissolved and transferred. The latter process is repeated.

A classical method of addition of liquids to a reaction is an addition funnel. Within a closed system it is necessary that the funnel be of the pressure-equalizing type (Fig. 5.15),

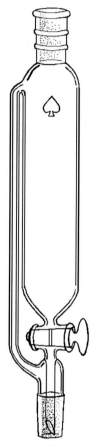

Figure 5.15 A pressure equalizing dropping funnel for addition under an inert atmosphere.

with a tube connecting the space above the reagent to the space above the addition tube but below the joint. A potential problem with addition funnels is that the addition rate is dependent not only on the angle of the stopcock but on the hydrostatic pressure of the solvent, meaning that as liquid is depleted, the addition rate slows. Maintaining a constant rate of addition using an addition funnel requires close monitoring that is tedious. Another important method of liquid addition is a syringe pump (Fig. 5.16), also called (often in experimental descriptions) a motor-driven syringe. This instrument works with a standard syringe (see Chapter 9, Section 9.4), basically clamping it in place and driving a ram against the plunger. The syringe pump can be set for a range of delivery rates for a given syringe size, and it is a superior tool to obtain a constant rate of addition.

It is worth specifically mentioning the purpose of varying the addition rate in a reaction. One obvious purpose is that the reaction is exothermic, and fast addition rates lead to thermal runaway. That is, the heat of reaction from addition of the first portion of the reagent raises the temperature of the reaction mixture such that the next portion of the reagent reacts even faster, creating even more heat that raises the reaction temperature even more.

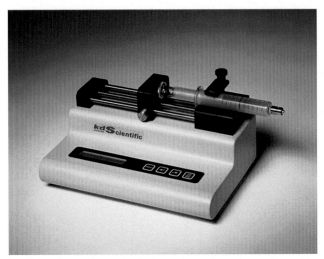

Figure 5.16 A syringe pump accepts a variety of sizes of syringes and drives a ram against the plunger at a constant rate. The rate of liquid delivery is variable based on the size of the syringe and the setting of the syringe pump.
Photo provided by KD Scientific.

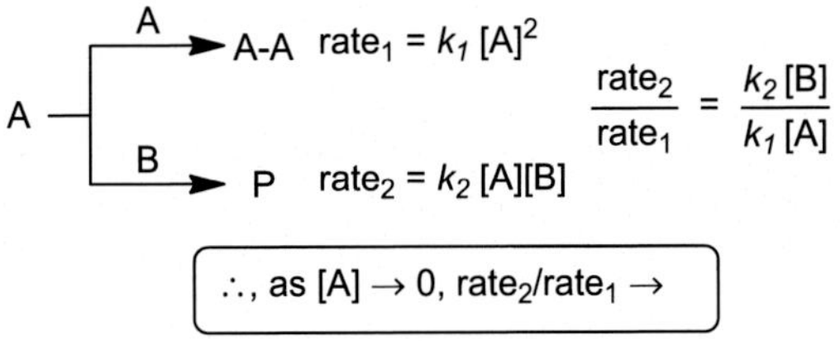

$$A \begin{cases} \xrightarrow{\;A\;} A\text{-}A \quad rate_1 = k_1\,[A]^2 \\ \\ \xrightarrow{\;B\;} P \quad rate_2 = k_2\,[A][B] \end{cases} \qquad \frac{rate_2}{rate_1} = \frac{k_2\,[B]}{k_1\,[A]}$$

$$\therefore, \text{ as } [A] \to 0,\ rate_2/rate_1 \to$$

Figure 5.17 A kinetic scheme that demonstrates how the mode of addition can affect the proportion of self-coupling versus cross-coupling in a reaction involving reagents A and B.

Slow addition of reagents is commonly used to minimize the reaction of a reagent with itself. The principle is as follows. We wish A to react with B to form the product P, but A can also react with itself. From the rate laws for each reaction, the relative rates of the two processes can be compared (Fig. 5.17). This analysis shows that the way to favor reaction with B (i.e., $rate_2$) is to keep the concentration of A as low as possible. This is true irrespective of the relative magnitude of the rate constants for the two reactions or the B concentration. A situation where the A concentration is as low as possible is easy to arrange by adding A slowly to the solution of B. All of the A that is added in each drop is ideally consumed by reaction with B before the next drop of A is added.

Slow addition to keep the reagent concentration low is also used to favor intramolecular reactions at the expense of intermolecular reactions. For example, the formation of large rings (Fig. 5.18, $n=7$ or greater) must compete with the formation of oligomeric derivatives. Writing the rate law for each reaction, taking their ratio, and solving the equation is an exercise left to the reader. This analysis shows that the way to favor intramolecular reaction (i.e., $rate_1$) is to keep the concentration of C as low as possible, which again can be achieved by a slow drop-wise addition to the reaction mixture.

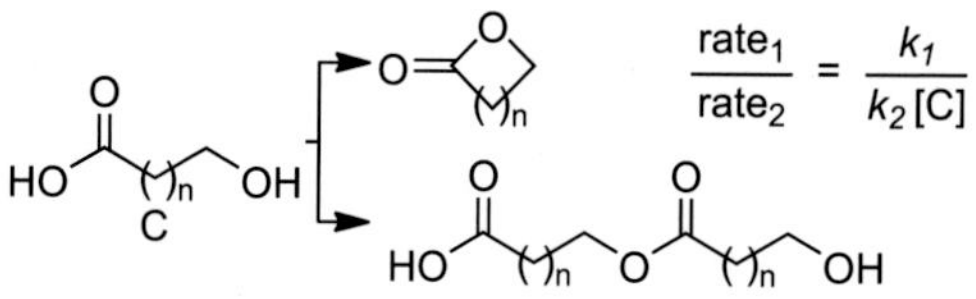

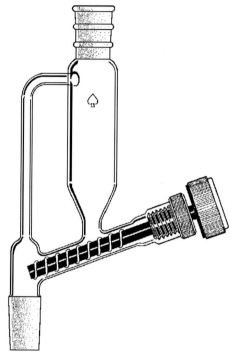

Figure 5.18 A kinetic analysis demonstrates that slow addition, maintaining a low concentration of reagent C, maximizes the proportion of intramolecular reaction to intermolecular reactions.

Figure 5.19 A powder addition funnel can be used only with free-flowing solids that do not adhere to glass. It also requires a relatively large-scale reaction.

A powder addition funnel (Fig. 5.19) can be used to add solids to a reaction mixture. Alternatively, tip flasks (with a bent neck) retain solids when in the down position and allow them to drop into the reaction mixture when rotated into the up position.

These techniques for working under an inert atmosphere will find the most use in ordinary organic laboratories. More sophisticated methods are also available, as reflected in the outstanding text of Shriver and Drezdzon (1986).

5.6 Condensers

When reactions are conducted under reflux, a condenser is used. Two basic types of condensers are available: water and dry ice (Figs. 5.20 and 5.21). The dry ice condenser is used for cryogenic solvents such as liquid ammonia. It is fitted to the flask and filled

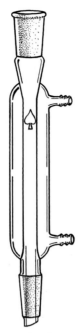

Figure 5.20 A conventional water-cooled condenser.

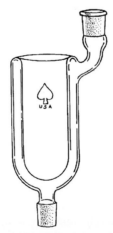

Figure 5.21 A dry ice condenser.

with dry ice/isopropanol. The chemist must attend this condenser closely to ensure that the dry ice is replenished when needed. Water condensers should be connected to a supply of cold water with Tygon tubing. It can greatly enhance the ability to connect varying combinations of condensers into the system if the waterline is fitted with Nalgene connectors. Each condenser is equipped with an inlet and outlet tube with a connector, allowing the easy insertion of a condenser into the system. The water flow in condensers should be carefully controlled with a needle valve or pressure regulator if reactions

Figure 5.22 A flow indicator for insertion into tubing that delivers cooling water to a condenser. The ball circles the device when water is flowing through the tubing. Bel-Art SP Scientific.

are to be left unattended (see Chapter 9, Section 9.8). Otherwise, the drop in pressure when water use rises may cause the reaction to lose cooling, or a rise in water pressure could cause a connection to loosen and lead to a flood. It is also a good idea to insert a flow indicator (Fig. 5.22) into the water cooling lines so that the speed of water flow is apparent. Floods from cooling water overflows are one of the most common laboratory accidents. While not as severe as some chemistry accidents, floods are still quite damaging, and the water damage can extend from the lab of the experimenter to labs in floors below. It is essential for this reason that all connections of water hoses to hose connectors on glassware be secured with copper wire. Automatic reaction monitors/controllers are also available that sense whether an apparatus is receiving cooling water and turn off both the water flow (with a solenoid) and the heat if cooling has been lost.

5.7 Other Equipment and Considerations

A jack stand or lab jack (Fig. 5.23), essentially a small scissors-type jack, is extremely useful for a reaction setup. It permits heating/cooling baths, magnetic stirrers, and/or hotplates to be brought up to (and quickly withdrawn from) a flask that is mounted at a fixed height to a fixed grid of metal rods. These grids are standard in most laboratory furnishings. Ring stands are not recommended for supporting reaction flasks because they are relatively unstable and easily toppled. The jack stand is also used to adjust the distance between a magnetic stirring motor and the stir bar in the flask to achieve optimal stirring.

Minimizing exposure of workers to chemicals is one of the hallmarks of laboratory safety. There is no justification for conducting reactions anyplace except in the fume hood. Most modern laboratories and standard procedures are set up with this in mind, with areas outside the hood reserved for other operations. It is crucial that all chemists be aware of proper fume hood operation, which should be an important element of the

Figure 5.23 A lab-jack or jack stand is very useful for raising and lowering temperature control baths to a flask that is held at a fixed position.

institutional safety training received. A common failing is leaving the sash too high, which reduces the efficiency of that specific hood and needlessly wastes resources in the exhaust system as a whole.

One example of how the principle of minimizing exposure can be important is with reagents that fall into the class of "sensitizers." Often these are strong alkylating agents. After an initial exposure of the chemist to such a reagent, future exposures to that compound may promote an allergic reaction such as a rash or an anaphylactic reaction such as asthma, and this sensitivity may extend to other compounds in that class. An initial exposure to one compound may prevent the chemist from ever again working with a whole range of reagents. Knowledge aforethought of this phenomenon can inform many laboratory procedures. For example, it is fairly easy to become sensitized to methylating agents. Knowing this, it would be unwise to quench a reaction involving excess tosyl chloride with methanol, which would be converted to methyl tosylate (Eq. 5.7), an effective methylating agent. A better choice for a quenching agent is ethanol, which is almost as reactive as methanol in quenching the tosyl chloride but generates ethyl tosylate, which is not nearly as reactive as an alkylating agent as methyl tosylate.

$$\text{MeOH} + \text{TsCl} \longrightarrow \text{Me-OTs} \tag{5.7}$$

Reference

Shriver, D.F., Drezdzon, M.A., 1986. The Manipulation of Air-Sensitive Compounds, second ed. Wiley, New York.

Temperature Control

6

Chapter Outline

Reaction temperatures commonly used in organic reactions range from $-100°C$ to nearly $200°C$. For reactions below room temperature, an alcohol thermometer will typically be used for temperature monitoring. For reactions above room temperature, a mercury thermometer should be used. Thermometer adapters made of Teflon or glass (Fig. 6.1) are available for standard sizes of ground-glass joints. The former have neoprene rings that can be tightened onto the thermometer via a screw fitting. The latter might work in this way, or have very small tapered ground-glass joints that accept thermometers with matching ground-glass fittings. The latter thermometers are less attractive because they are more expensive, and because their depth of immersion into the flask, which may not be correct for every reaction, is fixed. A tricky issue in getting a workable reaction setup is choosing the right flask and solvent volumes to allow the thermometer bulb to be immersed in the solution without being struck by the magnetic stirring bar, and to have this situation persist once the solution is being stirred, which often creates a vortex in the center of the solution just where the thermometer bulb sits.

6.1   Heating

Chemistry was revolutionized when Robert Bunsen perfected the gas burner, but open flames are an anachronism in today's laboratory. Heating is essential to many reactions, though, so several solutions will be discussed.

6.1.1   Conventional

For reactions above room temperature, oil baths provide the best control. Silicone oil, also called DOWTHERM fluid, heat transfer fluid, or just transfer fluid, is preferred. Many variants with different properties are available—they typically have a working temperature up to about $250°C$ (Fig. 6.2). Do not use paraffin, mineral, or white oils, since they smoke and discolor and have a low flash point.

Handbook of Synthetic Organic Chemistry. http://dx.doi.org/10.1016/B978-0-12-809504-1.00006-6

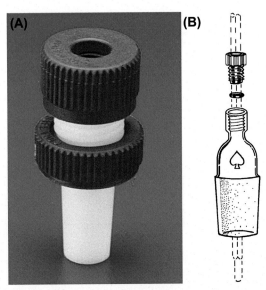

Figure 6.1 (A) A Teflon thermometer adapter fits into a standard ground-glass joint and accepts a glass thermometer or gas inlet tube. The seal is made by an O-ring that is compressed against the thermometer or tube by screwing in the knurled knob. Bel-Art - SP Scientific. (B) Thermometer adapters can also be made from glass.

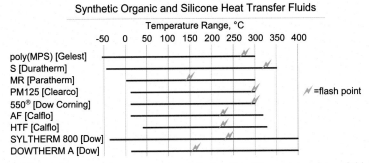

Figure 6.2 The working temperature ranges and flash points for several commercial heat transfer fluids.

Oil baths are best heated electrically by an integral heating element (Fig. 6.3) controlled by a Variac or Powerstat. Some labs have a tradition of using homemade oil bath heaters consisting of coils of nichrome wire connected to a Variac or Powerstat with patch cords. These setups typically have exposed electrical contacts and are not safe.

Oil baths might also be heated by a hotplate or stirring hotplate. This is probably the easiest and most common way used to heat a bath, but that does not mean it is the best. The temperature the bath can reach is limited by poor thermal contact between the top of the hotplate and the bath container, so do not use foil or insulating material on the top plate. Because the temperature of the hotplate must be higher, sometimes much higher, than the temperature of the bath, it presents a fire hazard when hotter than the flash point of the oil. Few hotplates are explosion-proof. Spills of oil or organic liquids

Figure 6.3 An oil bath with an integral heating element.

on the hot surface can also lead to a fire, or at least discoloration or failure of the ceramic top. The excess heat given off by the hotplate can create problems by heating the surrounding apparatus. It is important to use a thermometer in the bath itself so its temperature can be directly monitored and controlled. A magnetic stir bar should also be used in the oil bath to ensure its temperature is uniform. Electronic devices to monitor and maintain the oil bath temperature are available—one example is the THERM-O-WATCH (Fig. 6.4). It has a sensor that clamps onto a mercury thermometer to measure the mercury level, and turns the heat on and off as needed to maintain the mercury level and therefore the temperature. Of course, this cycling can result in drifting of the bath temperature by ±10°C. When filling an oil bath, be sure to take into account the volume displaced by the flask and volume expansion upon heating.

Heating mantles are really only satisfactory for heating reactions conducted at reflux, and even then there is a risk of uneven heat distribution across the mantle. Take care that reflux is gentle, ensuring the pot is not appreciably above reflux temperatures. It can be difficult to get magnetic stirring to work through a heating mantle, however. Heating mantles may be used for solvent stills. Steam baths may be used to heat reactions to temperatures in the 70°C range but are hardly available in the modern organic chemistry laboratory. Neither of these methods has much to recommend it for reaction heating.

6.1.2 Microwave

A modern approach to the heating of organic reaction mixtures uses microwaves. For this purpose, the use of specialized microwave equipment made for laboratories (Fig. 6.5) is essential. Not only are conventional domestic microwave ovens hazardous to use for chemical processes, most respected journals do not allow experimental descriptions that use them. Microwave chemistry has become a prominent synthetic organic subfield, and texts are available that describe its many nuances in great detail (Kappe et al., 2009, 2012).

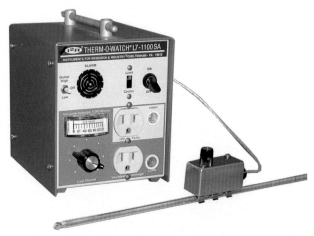

Figure 6.4 A THERM-O-WATCH monitors the temperature via a sensor placed on a mercury thermometer and controls the power being delivered to a heating device to maintain a constant temperature. Similar sensors are available to ensure water is flowing before heat is applied to a system that requires cooling.

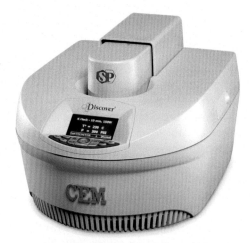

Figure 6.5 A microwave reactor for use in organic synthesis.

Microwaves heat materials by dielectric and/or ionic conduction mechanisms. In the first, the electric component of electromagnetic radiation interacts with the dipoles in solvent molecules, inducing them to oscillate with the electric field and causing heating through molecular motion, friction, and dielectric loss. Several molecular properties are related to these heating mechanisms, but there is no simple, intuitive measure that indicates a solvent's capacity for microwave heating. The bulk properties of the liquid govern this parameter, which is measured by the dielectric loss tangent ($\tan \delta$). Values for $\tan \delta$ are compiled for a range of common solvents in Table 6.1. Higher loss tangents indicate solvents more readily heated by microwaves. Given the general familiarity with

Table 6.1 Loss Tangents (20°C) for Select Synthetic Solvents Used in Microwave Reactions

Solvent	tan δ	Solvent	tan δ
Ethanol	0.941	Dimethylformamide	0.161
Dimethylsulfoxide	0.825	Water	0.123
Isopropanol	0.799	Chlorobenzene	0.101
Formic acid	0.722	Acetonitrile	0.062
Methanol	0.659	Ethyl acetate	0.059
1-Butanol	0.571	Acetone	0.054
2-Butanol	0.447	Tetrahydrofuran	0.047
o-Dichlorobenzene	0.280	Dichloromethane	0.042
N-methyl-2-pyrollidone	0.275	Toluene	0.040
Acetic acid	0.174	Hexane	0.020

domestic microwave heating via the water in foodstuffs, it is surprising that water has a smaller loss tangent than several organic solvents. Solvents with a tan δ greater than water are certainly useful for organic reactions heated by microwave radiation. Solvents with very low tan δ also have little to no dipole moment; they are essentially transparent to microwaves and cannot be heated by themselves . However, in real reaction mixtures, reactants and reagents that *are* microwave-active may be present and facilitate heating, or solvents/additives can be included as antennas to enable microwave heating.

For the ionic conduction mechanism, microwaves cause ions in solution to oscillate with the electric field, heating the liquid in ways similar to the dielectric loss mechanism. This pathway is available for aqueous solutions of ions, organic ionic liquids, and organic liquids in which (hydrophobic) ionic materials are dissolved (e.g., phase transfer catalysts such as tetraalkylammonium salts).

Commercial microwave reactors provide capabilities to monitor and control the temperature, microwave power, and pressure (in closed reaction vessels; reactions can be conducted above atmospheric pressure). They have explosion-proof reaction/irradiation cavities that enable the use of organic solvents. Reactors often use infrared sensors in the microwave reaction cavity for temperature monitoring of the reaction vessel. This capability is crucial, since the internal temperature is essential data to document an experiment (as well as to publish microwave reactions in leading journals). Reactors can also include magnetic stirring, an external coolant gas for use during a reaction run or after it is complete, and computer control of reactor functions. The reactor should be located in a fume hood.

Commercial microwave reactors fall into two broad classes, called multimode and monomode. The former has some parallels to a conventional microwave oven: microwaves are reflected by the walls of the cavity and a mode stirrer is used to make their distribution more uniform. They can accommodate multiple reaction vessels, permitting different microwave reactions or replicate single reactions to be conducted in parallel. Reaction vessels are rotated on a rotor to ensure uniformity of heating. Multimode reactors are more scalable.

Monomode reactors direct radiation from the microwave source to a single reaction vessel, creating a standing wave with the sample positioned at the point of maximum energy. Vessels can be closed, special glass test tubes of 10–80 mL, sealed with a Teflon-coated septum and a plastic cap. A pressure-sensing head locks onto the tube and penetrates the septum with a fine needle to read the pressure. Open vessels at atmospheric pressure can also be accommodated, using glassware with capacity up to 125 mL, a conventional condenser, and provision for inert gas. Solvent volumes should be about half of vessel volume.

For some reactions, microwave heating yields far superior results to conventional heating, and much inquiry has been made into why this should be so. One consideration familiar from domestic microwave ovens is that the reaction is heated throughout, without the need for heat transfer from the exterior by convection. Infrared imaging of organic reactions during microwave heating has also shown uneven thermal distribution. Yet, the external vessel wall is *cooler* than the interior solution, opposite to the situation with conventional heating. It is therefore still important to stir solutions in a microwave reactor to heat them uniformly.

For any microwave reaction, parameters that can be controlled/limited include microwave power (in watts), temperature, and pressure (in psi) if a sealed vessel is used. A typical experimental run is initiated in the reactor control software, where the solvent, reaction time, and reaction temperature are specified. Limits on power, pressure, and temperature can also be entered. The reactor will stop the run if these limits are exceeded. When reaction is initiated, the reactor will use the maximum permitted power to bring the vessel to the set temperature. It records the temperature and pressure throughout the reaction time and provides a profile for each, which can be important documentation for experimental descriptions.

The efficacy of microwave chemistry is truly incomparable to any other heating technique. Heating rates are such that a moderately microwave-active solvent such as N-methylpyrrolidone can be heated to 120°C in 1 min using 300 W of power. Solvents can be superheated to well above their boiling points, raising the pressure in the reaction vessel substantially (some have pressure limits of 200 psi). Superheating can be substantial, such as methanol heated to 195°C, something that could not be easily achieved with conventional heating. In fact, this capability has affected the way workers perform microwave reactions. Examining Table 6.1, many of the solvents with the largest $\tan \delta$ also have relatively high boiling points, making them difficult to remove following the reaction. By conducting reactions in a sealed vessel with a volatile microwave-active solvent heated to far above its boiling point, the need for high boiling solvents is eliminated.

A major advantage of microwave heating seems to be the speed with which the reaction mixture reaches operating temperature and with which it can be cooled following reactions. To some extent the best analogy of microwave heating in conventional organic reaction techniques may be flow pyrolysis, a method that has a good history in organic synthesis but is operationally difficult to implement and therefore not discussed in this text. There is also some analogy of microwave heating to processes conducted in sealed tubes (Chapter 9, Section 9.7.2), as microwave reactors have pressure limits of 20–30 bar. That section can be consulted regarding the pressures to be expected when sealed reaction systems are heated.

6.2   Cooling

For low temperature, several different liquid baths are used with the few available coolants: ice, dry ice, and liquid nitrogen. Temperatures of various baths are given in Table 6.2. However, temperatures around the freezing points of the coolants themselves are far easier to maintain. At the other temperatures, the chemist is often relying on the

Table 6.2 **Coolant/Liquid Combinations to Achieve Specified Coolant Bath Temperatures**

Temperature (°C)	Coolant/Liquid
4 to 0	Ice/H_2O
−10 to −15	Ice/acetone
−10 to −15	100 g Ice/33 g NaCl
−16	100 g Ice/25 g NH_4Cl
−28	100 g Ice/67 g NaBr
−34	100 g Ice/84 g $MgCl_2 \cdot 6H_2O$
−55	100 g Ice/143 g $CaCl_2 \cdot 6H_2O$
−10.5	Dry ice/ethylene glycol
−12	Dry ice/cycloheptane
−15	Dry ice/benzyl alcohol
−25	Dry ice/1,3-dichlorobenezene
−29	Dry ice/o-xylene
−30 to −45	Dry ice/aq. $CaCl_2$, varying concentration
−32	Dry ice/m-toluidine
−38	Dry ice/3-heptanone
−41	Dry ice/acetonitrile
−46	Dry ice/cyclohexanone
−47	Dry ice/m-xylene
−56	Dry ice/n-octane
−78	Dry ice/commercial isopropanol[a]
−83.6	Liquid N_2/ethyl acetate
−89	Liquid N_2/n-butanol
−94	Liquid N_2/hexane
−94.6	Liquid N_2/acetone
−95.1	Liquid N_2/toluene
−98	Liquid N_2/methanol
−104	Liquid N_2/cyclohexane
−116	Liquid N_2/ethanol
−120	Liquid N_2/4:1:1 petroleum ether/isopropanol/acetone
−131	Liquid N_2/n-pentane
−160	Liquid N_2/isopentane

[a]Dry ice/isopropanol is recommended over dry ice/acetone because it fizzes less, is less volatile, and is less flammable.

Figure 6.6 A homemade insulated cold bath.

freezing point of the liquid to buffer the effect of added coolant. This process seems to degenerate into incessant addition of coolant and then liquid, which is not optimal, since it is certain that the reaction temperature is actually fluctuating around the melting point of the liquid (not to mention the tedium imposed on the chemist). This problem may be addressed by using eutectic ice/salt mixtures of specific proportions. When temperature maintenance for long periods is needed, homemade insulated baths consisting of one crystallizing dish inside another, with a piece of Tygon tubing as a spacer between them and the space between filled with a packing material or vermiculite and sealed with silicone sealer (Fig. 6.6), are sometimes used to conserve coolant. This is a cheap alternative to a dewar dish (Fig. 6.7), and also gets around the problem that many dewars prevent effective operation of magnetic stirrers.

Liquid nitrogen (LN) is one of the few cryogenic liquids routinely used in the organic chemistry laboratory, and special considerations attend its use. Obviously it can freeze flesh, so it must be used with caution and appropriate personal protective equipment. LN is transferred from large storage dewars (Fig. 6.8) into smaller dewars (Fig. 6.9) by simply opening the outlet of the former. Cold nitrogen gas will initially flow into the receiver, to be followed by liquid once the transfer line has been cooled below the boiling point of nitrogen. Then liquid reaching the receiver will be vaporized in cooling it below the boiling point of nitrogen, and finally liquid will begin to accumulate in the receiver. This process repeats with every transfer into a new container.

Chillers offered by several different manufacturers (Fig. 6.10) use a cooling probe in place of a solid coolant and, most usefully, have a temperature probe so that a thermostat can maintain the temperature at a set-point. These chillers can be essential to maintaining reactions at low temperature for longer than the chemist's stamina and patience. A cooling liquid commensurate with the desired temperature (see Table 6.2) must be used.

Figure 6.7 Low-profile dewar flasks for use as cold baths.

Figure 6.8 Large dewars for storage of liquid nitrogen. These tanks are around 1.5 m in height.

Figure 6.9 Smaller dewar container used to transfer liquid nitrogen from the central supply dewar to the individual lab.
© Sigma–Aldrich Co. LLC.

Figure 6.10 A chiller with a probe that can be used in place of a coolant.
Photo provided by Lauda-Brinkmann.

References

Kappe, C.O., Stadler, A., Dallinger, D., 2012. Microwaves in Organic and Medicinal Chemistry, second ed. Wiley-VCH, Weinheim.

Kappe, C.O., Dallinger, D., Murphree, S.S., 2009. Practical Microwave Synthesis for Organic Chemists. Wiley-VCH, Weinheim.

Solvents

7

Chapter Outline

7.1   Selection

The selection of a solvent for today's reaction may be obvious because a procedure is available for this or a comparable reaction. A detailed consideration of solvent effects on organic reactions is not intended here; excellent reference texts are available (Reichardt and Welton, 2010). In addition there are some statistical methods available that attempt to segregate solvents based on aggregates of their molecular properties (Carlson et al., 1985). A chart of these solvent properties is supplied in Appendix 2, as well as suggestions on how to use the chart to select solvents. It offers a way to consider two main solvent traits, polarity and polarizability, by simple visual retrieval. Appendices 2–5 address several other issues that can arise with solvents for organic reactions, and they should be consulted before choosing one.

In selecting a solvent, knowledge of the risk of exposure of the chemist to the solvent must be taken into account. The relative risk involved with particular solvents, and therefore the precautions that should be observed, may not always be obvious. Some criteria are discussed in Appendix 5. However, as has been mentioned earlier, a key principle of chemical hygiene is to minimize exposure of the chemist to any chemical. Precautions appropriate to the particular solvent, such as ventilator use for inhalation toxicity, or gloves, lab coats, and other personal protective equipment for skin toxicity, should be observed.

Another criterion for solvent selection that has arisen recently is variously called sustainability, environmental impact, or greenness. The Pharmaceutical Round Table of the American Chemical Society Green Chemistry Institute has been a prime mover to consider this topic, so the rankings have a significant industrial leaning. Different groups have developed their own scoring criteria for green solvents. One example is shown in Fig. 7.1 (Prat et al., 2013). This classification was based on safety, occupational health impacts, environmental impacts, cost, and recyclability.

Handbook of Synthetic Organic Chemistry. http://dx.doi.org/10.1016/B978-0-12-809504-1.00007-8

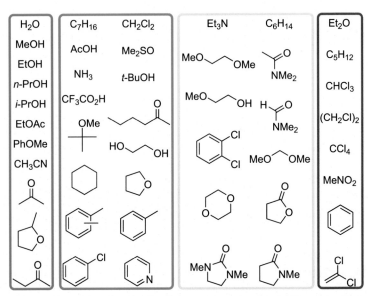

Figure 7.1 Organic solvents classified for greenness into four bands. The green (dark gray in print versions) *box* encompasses preferred solvents. The yellow-green (light gray in print versions) *box* is solvents for which greener substitutes are suggested. The yellow (very light gray in print versions) *box* is solvents for which greener substitutes should be used. The red (gray in print versions) *box* is solvents ranked lowest in greenness that should not be used. Classification from Prat, D., Pardigon, O., Flemming, H.-W., Letestu, S., Ducandas, V., Isnard, P., Guntrum, E., Senac, T., Ruisseau, S., Cruciani, P., Hosek, P., 2013. Sanofi's solvent selection guide: a step toward more sustainable processes. Org. Proc. Res. Dev. 17, 1517–1525. doi:10.1021/op4002565.

Modern replacements for problematic traditional solvents are continually being developed. Much old literature used benzene or carbon tetrachloride, as they were quite available to early chemists, but are regarded today as toxic and undesirable (see Appendix 5). Replicating a procedure that originally used them calls for a substitution. One solvent in particular that chemists have had a hard time giving up is dichloromethane; it has a long history and many desirable features, but its health effects are now in question. An alternative is benzotrifluoride or trifluorotoluene (Maul et al., 1999), which can substitute for dichloromethane as a reaction solvent for Lewis acid reactions, oxidations, and radical reactions. It is also a logical replacement for benzene or carbon tetrachloride, but is not recommended for extractions.

Solvent substitutions for classic ethereal solvents such as diethyl ether and THF have also been contemplated. Dimethoxymethane and *tert*-butyl methyl ether have been proposed as safer replacements. An alternative to THF is 2-methyltetrahydrofuran. A key difference from THF is that it is not miscible with water, making extraction steps simpler. The workup methods offered for water-miscible solvents (Chapter 11) are not required.

Chemists often aim to address multiple properties desired in a reaction solvent by using solvent mixtures. However, sometimes it is discovered only upon actually mixing two solvents that they are not miscible. Solvent miscibility is covered in Appendix

3. Also be sure to consult freezing points of potential reaction solvents (Appendix 4) before choosing a low temperature for a reaction. These types of missteps are often made with solvents such as DMSO and acetic acid that are naturally thought of as liquids but freeze/melt just below room temperature. Solvent mixtures have lower freezing points and may allow access to lower temperatures, however. For example, 4:4:1 tetrahydrofuran/ethyl ether/pentane (the so-called Trapp solvent) is useful to −110°C. Some solvents naturally are mixtures. Reference to petroleum ether or Skelly B is sometimes seen in older literature. This is a petroleum distillation fraction (35–60°C) composed of C5–C6 aliphatic hydrocarbons. It presumably got its name because of volatility similar to ethyl ether. It is chemically similar to the solvent hexanes, a mixture of C6 hydrocarbons dominated by n-hexane and methylcyclopentane, but petroleum ether might include unsaturated hydrocarbons.

7.2   Purity

As the component of the reaction mixture present in the largest amount, solvent purity can have a major impact on a reaction. Of particular concern is the removal of dissolved water or other protic substances from solvents. Water is pervasive and therefore will be present in any solvent with an affinity for it. This includes essentially all ethers and dipolar aprotic solvents; it is less of a problem with hydrocarbons. Not only will dissolved water serve as an acid for any of the interesting organic anions frequently used in synthesis (Eq. 7.1), water interferes with the formation of organometallic reagents such as Grignard's from alkyl halides. Solvent purification typically has involved distillation from a drying agent appropriate for each solvent. A classical choice for ethers has been sodium metal, often in the presence of an indicator such as benzophenone. Sodium's reducing properties address the very real concern with peroxides that can be formed from ethers in the presence of oxygen. However, as mentioned later, alkali metal stills are now considered too hazardous for routine use and alternative purification methods are used.

$$\text{Li}\diagdown\diagup\text{CN} + \text{H}_2\text{O} \longrightarrow \text{CH}_3\text{CN} + \text{LiOH} \tag{7.1}$$

A surprisingly large number of organic compounds react spontaneously with O_2 in the air to form peroxides. The ease with which this occurs by structural class is given in Fig. 7.2 (Kelly, 1996). Diisopropyl ether (and likely other ethers with tertiary α-hydrogens) can form explosive levels of peroxides even without concentration by evaporation or distillation. Butadiene also forms peroxides at explosive levels when stored as a neat liquid. This may not seem very threatening since butadiene is typically used as a gas or generated in situ. However, the same admonition likely applies to many other dienes, and they certainly are stored in neat form. A large number of compounds spontaneously form peroxides that can reach explosive levels upon concentration (since peroxides are less volatile than their precursors). They include acetaldehyde, benzyl alcohol, 2-butanol, cumene, cyclohexanol, cyclohexene, 2-cyclohexen-1-ol, decahydronaphthalene, diacetylene, dicyclopentadiene, diethyl

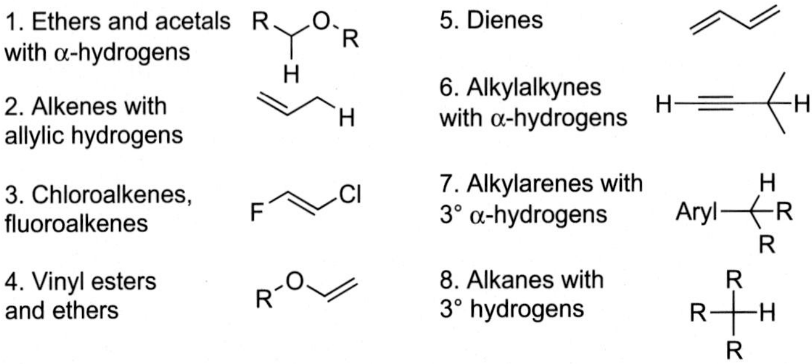

1. Ethers and acetals with α-hydrogens

2. Alkenes with allylic hydrogens

3. Chloroalkenes, fluoroalkenes

4. Vinyl esters and ethers

5. Dienes

6. Alkylalkynes with α-hydrogens

7. Alkylarenes with 3° α-hydrogens

8. Alkanes with 3° hydrogens

Figure 7.2 Peroxidizable organic structures in order of decreasing reactivity with O_2 from 1 to 8.

ether, diglyme, dioxanes, glyme, 4-heptanol, 2-hexanol, methylacetylene, 3-methyl-1-butanol, methylcyclopentane, methyl isobutyl ketone, 4-methyl-2-pentanol, 2-pentanol, 4-penten-1-ol, 1-phenylethanol, 2-phenylethanol, 2-propanol, tetrahydrofuran, and tetrahydronaphthalene. Of the commonly used laboratory ethers, tetrahydrofuran likely is peroxidized fastest.

Peroxide-forming chemicals should be stored in the original manufacturer's container whenever possible. This is very important in the case of diethyl ether because the iron in the steel containers in which it is shipped acts as a peroxide inhibitor. The date it was opened must be recorded on each container. In general, peroxide-forming chemicals should be stored in sealed, air-impermeable containers and should be kept away from light, which can initiate peroxide formation. Dark amber glass with a tight-fitting cap is recommended. If at all possible, they should be placed under an inert atmosphere. Peroxide-forming chemicals can be stored for 3 months after opening the container if they form peroxides without concentration, and for 12 months after opening the containers if they form peroxides with concentration. Materials may be retained beyond this suggested shelf life only if they have been tested for peroxides (see later), show peroxide concentrations lower than 100 ppm, and are retested frequently.

Safety Note

Researchers should never test containers of unknown age or origin for peroxides. Older containers are far more likely to have concentrated peroxides or peroxide crystallization in the cap threads. Therefore they can present a serious hazard when opened for testing.

All solvents that are to be distilled should be tested for the presence of peroxides regardless of how new they might be. A safe level for peroxides is considered to be less than 100 ppm. While several methods are available to test for peroxides in the laboratory, the most convenient is the use of peroxide test strips available from many

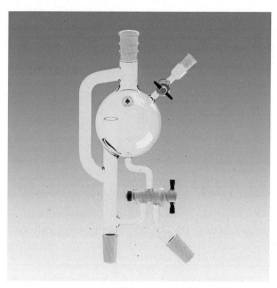

Figure 7.3 A still head for a solvent still.

chemical suppliers. For volatile organic chemicals, the test strip is immersed in the chemical for 1 s. The chemist breathes on the strip for 15–30 s or until the color stabilizes, and the color is compared with a provided colorimetric scale. Any container found to have a peroxide concentration over 100 ppm should be disposed of (with the assistance of the safety office).

Solvents stills are traditionally a continually maintained fixture in many organic synthesis laboratories. However, still pots containing highly flammable solvents and reactive metals pose a significant fire risk, not only during use but during quenching when the still must be regenerated. This book will therefore not provide a detailed procedure for setting up a sodium solvent still. It makes more sense to maintain a still system continuously when other, less dangerous drying agents are used and the lab uses that solvent regularly. An apparatus for that purpose collects distilled solvent in a bulb (Fig. 7.3). Stills are usually maintained at a low heating level (not enough to cause reflux) so that they quickly come to reflux to supply needed solvent. After collecting solvent from a still and turning the heat back down, the nitrogen flow must be maintained at a sufficient rate so as to avoid pulling oil back into the system from the bubbler (recall the gas laws).

Available drying agents other than sodium include those that react with water chemically (Eqs. 7.2–7.4). The desiccant(s) recommended for some commonly used solvents are given in Table 7.1. Drying agents that sequester water, such as molecular sieves, are useful with essentially any solvent given in Table 7.1. Sieves are zeolites (alkali aluminum silicates) with specific pore sizes; for example, 3 Å molecular sieves permit only water to penetrate into the solid. Molecular sieves are amazingly effective and broadly useful desiccants provided they have been activated, which is done by heating them under vacuum, ideally to 300–350°C, but heating to 150–200°C is still effective. They must be stored under an anhydrous atmosphere, like any desiccant. The properties of several classes of molecular sieves are given in Table 7.2. They are available in a variety of physical forms, including beads, pellets, and powders.

$$H_2O + P_2O_5 \longrightarrow \quad\quad\quad\quad\quad\quad\quad (7.2)$$

$$H_2O + CaH_2 \longrightarrow CaO + 2\,H_2 \quad\quad\quad\quad\quad (7.3)$$

$$H_2O + Ac_2O \longrightarrow 2\,HOAc \quad\quad\quad\quad\quad\quad (7.4)$$

Table 7.1 Desiccant(s) and Purification Method for Some Common Solvents

Solvent	Drying Agents	Fraction/Pressure/Other Info
Acetone	CaSO$_4$	Or stand over molecular sieves (24 h)
Acetonitrile	P$_2$O$_5$, CaH$_2$	
tert-Butanol	CaH$_2$, Al/Hg, Mg(OEt)$_2$	Melting point of 25°C means solid may block condenser
Dichloromethane	CaH$_2$, P$_2$O$_5$	Or stand over molecular sieves (24 h)
Diisopropylamine	CaH$_2$	
Dimethylformamide	BaO, P$_2$O$_5$	Stir with CaH$_2$, filter, and distill in vacuo[a] from the desiccant (55°C, 20 torr)[b]
Dimethylsulfoxide	CaH$_2$	Stand over CaH$_2$ overnight, distill in vacuo[a] (72°C, 12 torr), discard forerun, store over molecular sieves (4 Å)
Ethanol	Mg(OEt)$_2$, CaH$_2$	From Mg metal
Ethyl acetate	P$_2$O$_5$, Ac$_2$O, K$_2$CO$_3$	Collect all distillate
Hexamethylphosphoramide	CaH$_2$	Distill in vacuo[a] (115°C, 15 torr), store over molecular sieves (4 Å)
Hexanes	CaH$_2$	Collect all distillate
Nitromethane	mol sieves, CaSO$_4$	Collect all distillate
Methanol	Mg(OMe)$_2$, CaH$_2$	From Mg metal
N-Methylpyrrolidone	BaO, CaH$_2$	Distill in vacuo[a] (96°C, 24 torr)
Pentane	CaH$_2$	Collect all distilling below 55°C
Petroleum ether	CaH$_2$	Collect all distilling below 55°C
Pyridine	BaO, CaH$_2$	Collect all distillate
Toluene	CaH$_2$, P$_2$O$_5$	Collect all distillate
Trichloroethylene	K$_2$CO$_3$	Collect all distillate
Triethylamine	CaH$_2$	Collect all distillate

[a]Aspirator vacuum is usually effective. Be certain to insert a drying tube into the vacuum line.
[b]Decomposes at boiling point at atmospheric pressure.

Table 7.2 **Types of Molecular Sieves**

Type	Pore Size (Å)	Description
3A	3	Absorbs H_2O; good for drying polar liquids.
4A	4	Absorbs H_2O, C_2H_5OH, C_2H_4, C_2H_6, C_3H_6 (but not higher hydrocarbons); good for drying nonpolar liquids and gases.
5A	5	Absorbs n-C_4H_9OH and n-C_4H_{10}, but not iso-alkanes or rings ≥C4.
10X	8	Absorbs highly branched hydrocarbons and aromatics; used for purification and drying of gases.
13X	10	Particularly good for drying hexamethylphosphoramide, $[(CH_3)_2N]_3PO$.

It is notable that CaH_2 is often used as a drying agent, generating hydrogen gas and CaO. It is attractive because it reacts readily but in a controlled way with water, unlike some of the more reactive hydride sources that can be dangerous. However, its broad use begs the question of what to do with a pot residue after the dried solvent has been distilled out. Residual CaH_2 can be decomposed by adding 25 mL of methanol for each gram of hydride, with stirring. Once reaction ceases, an equal volume of water is carefully added to the stirring slurry. The mixture is then neutralized with acid and disposed of as chemical waste.

The effectiveness of a range of desiccants as drying agents for common solvents (without distillation) has been examined analytically (Williams and Lawton, 2010). Alumina, silica, and 3 Å molecular sieves proved the best, being capable of reducing residual water content in a solvent to as low as 1 ppm for hydrocarbons and 6 ppm for polar solvents.

A contemporary solution to the problem of generating pure, anhydrous solvents without a still was provided by Grubbs (Pangborn et al., 1996). This method relies on a source of purified bulk solvents that is suited to using cartridges of highly activated desiccants to dry them. Depending on the solvent, these desiccants might include activated alumina or molecular sieves. Cartridges of supported copper catalyst are optionally used for removal of dissolved oxygen from hydrocarbons. Complete solvent delivery systems (SDSs) that enable the drying, manipulation, and supply of these solvents under inert gas are available (Fig. 7.4). Solvents can even be piped from the system directly into inert atmosphere glove boxes. The range of solvents that can be purified is wide, including acetonitrile, methanol, DMF, ethyl ether, tetrahydrofuran, hexanes, dimethoxyethane, toluene, pyridine, triethylamine, DMSO, and dichloromethane. Nitromethane cannot be used with these systems, and for alcohols, a glass insert in the cartridge is necessary to prevent leaching of desiccant into the solvent. Solvent systems are constructed to purify a particular number of solvents, ranging from three to a dozen or more, depending on the setting.

The special solvents used in an SDS are purchased in kegs similar to those used for beverages, typically of 20 L capacity. These kegs have both a solvent outlet and an inlet for inert gas to replace the volume of solvent withdrawn. There is no external

Figure 7.4 A "Grubbs" solvent purification system.
Photo provided by Inert.

indication of the amount of solvent remaining in a keg. It is therefore *crucial* that all users record in a logbook the amount of solvent they remove from the system each and every time they do so. When the amount of solvent remaining in a keg drops to less than 2 L, it must be replaced (since omissions in recording withdrawals likely have reduced it even farther).

The backbone of an SDS is a metal double manifold system for vacuum and argon. The vacuum is provided by a special oil-free diaphragm pump that is insensitive to volatile organic solvents. It is most convenient that the pump have a pressure-sensing mechanism that starts it whenever the pressure exceeds a set-point and stops it when a desired vacuum is reached.

The receiving bulb for each solvent is maintained under Ar when not in use, but when solvent is to be dispensed, the bulb is exposed to vacuum. Once the pressure reaches the set-point, the bulb is isolated from the vacuum (this step is crucial; this *must* be done before solvent is transferred). The bulb at reduced pressure can then be exposed to a solvent transfer line. The solvent is drawn from its keg, through a cylinder of desiccant appropriate for the solvent (previously set up on the system), and delivered into the bulb. It is important to deliver only the amount of solvent to be used. In classical solvent stills, the receiver bulb could be filled with solvent and whatever was not used could simply be allowed to drain back into the still pot. This possibility is excluded by the plumbing of the SDS; whatever collects in the bulb is removed from the system. Once the bulb contains the desired amount of solvent, the transfer line is switched off and the bulb is refilled with argon to atmospheric pressure. The solvent can then be removed from the bulb in a variety of ways, through glass fittings or via syringe or other transfers (as described in Sections 9.4 and 9.5).

After a sufficient volume of a particular solvent has been purified by the SDS, the desiccant in its cartridge will be discharged and the column will need to be replaced (either just the packing or the whole assembly). This is an involved process that is specific to each SDS vendor and will not be described in detail here.

7.3 Degassing

It may be important for some reactions to remove dissolved oxygen from the solvent. If this is not intrinsic to the solvent purification method itself (see earlier), solvent degassing can be performed in one of two ways. The rigorous procedure is called freeze–pump–thaw. The solvent is frozen well below its freezing point with a coolant, which for some solvents could be dry ice/isopropanol, and for others could be liquid nitrogen. A reasonably good vacuum (c. 1 torr) is applied for several minutes, and the container is refilled with inert gas. The solvent is thawed, and this cycle is repeated two to three times. The easier procedure that most synthetic chemists will use is to bubble an inert gas through the solvent for about 15 min using a fritted gas dispersion tube (Fig. 7.5). This method may not work well with particularly volatile solvents, which will evaporate.

Each solvent has its own unique properties that the chemist using it routinely should become familiar with. For example, chloroform has a tendency to undergo a slow decomposition to give HCl. This process is inhibited by the presence of ethanol, which is why commercial chloroform often includes low percentages of ethanol as a stabilizer. Other vendors use a compound called amylene for this purpose. The potential presence of ethanol, amylene, or HCl in chloroform must therefore always be considered.

Safety Note

When solvents are transferred in significant quantity from and to a metal or other conductive container, there is a risk of static electricity buildup between the reservoir and the receiver. If not neutralized, the static spark can ignite solvents with a low flash point, particularly ethers and hydrocarbons. These concerns can be addressed by bonding the metal containers to one another through conductive leads with alligator clips, as well as grounding one container. Metal safety solvent cans commonly found in labs do require bonding and grounding if dispensing into a metal container.

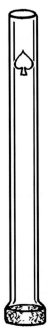

Figure 7.5 A gas dispersion or sparging tube has a glass frit at the end that creates voluminous bubbles.

7.4 Ammonia

A very special example of a reaction solvent is liquid ammonia. It is most commonly used for dissolving metal reductions, e.g., of unsaturated carbonyl compounds or aromatic rings. Liquid ammonia is obtained in a tank (see Section 4.2) that is connected to an apparatus including a charged dry ice condenser and a flask cooled in a −78°C bath. The tank valve is opened, ammonia is condensed into the flask, and the ammonia line is replaced by an inert gas line. The ammonia in the tank is not necessarily anhydrous, but for many reactions where excess sodium may be used, for example, cleavage of a benzyl ether using sodium in ammonia (Eq. 7.5), this may not be a problem.

$$\text{Ph}\diagdown\text{O}\diagdown\diagup\diagdown\text{CO}_2\text{Na} \xrightarrow[\text{NH}_3 \text{ (l)}]{\text{Na}^\circ} \text{HO}\diagdown\diagup\diagdown\text{CO}_2\text{Na} \qquad (7.5)$$

Sodium is packaged in a variety of ways and often appears as a large ingot, caked with NaOH from reaction with moisture in the atmosphere. Immediately upon opening the packaging, the sodium should be placed under mineral oil or a high-boiling hydrocarbon solvent such as xylene to protect it from the atmosphere.

The ductile metal is cut into small chunks (maximum 2 cm square) with a knife or spatula while under the mineral oil. The chunks are transferred with forceps to a container of toluene, which is used to remove the oil, and then to a container of the reaction solvent, which is used to remove the toluene. The chunks can also be briefly dipped at an intermediate stage into a container of dry alcohol, which will react slowly with the sodium to form hydrogen, cleaning its surface for the reaction. If the solvent is not too volatile, a beaker of it can be placed on a balance to weigh the sodium chunks.

Small pieces of sodium are added to the cooled ammonia. Sodium reacts with any water present to form sodium hydroxide (Eq. 7.6). Once all the water has been consumed, addition of more sodium will result in the formation of the unmistakable blue solution of solvated electrons (Eq. 7.7). The stoichiometric quantity of sodium needed for the reaction can now be added.

Alternatively, if the reaction in question requires that no hydroxide be present (e.g., a reductive alkylation of an enone; Eq. 7.8), the ammonia can first be condensed into a cooled round-bottom flask to which sodium is added until the blue color persists. This flask is then connected to the gas inlet of a dry ice condenser installed on the reaction apparatus. Anhydrous ammonia is distilled into this apparatus by slightly and carefully warming the flask with a water bath. An easier but less sure approach to obtain anhydrous ammonia is to pass it through two drying towers containing KOH and CaO before being condensed into the apparatus.

$$Na° + H_2O \longrightarrow H_2 + NaOH \tag{7.6}$$

$$Na° + NH_3 \longrightarrow e^-(NH_3)_4 \, Na^{\oplus} \tag{7.7}$$

$$\begin{array}{c} \text{1. Na° / NH}_3 \text{ (l)} \\ \xrightarrow{\hspace{2cm}} \\ \text{2. MeI / THF} \end{array} \tag{7.8}$$

References

Carlson, R., Lunstedt, T., Albano, C., 1985. Screening of suitable solvents in organic synthesis. Strategies for solvent selection. Acta Chem. Scand. B 39, 79–91. http://dx.doi.org/10.3891/acta.chem.scand.39b-0079.

Kelly, R.J., 1996. Review of safety guidelines for peroxidizable organic chemicals. Chem. Health Saf. 3, 28–36.

Maul, J.J., Ostrowski, P.J., Ublacker, G.A., Linclau, B., Curran, D.P., 1999. Benzotrifluoride and derivatives: useful solvents for organic synthesis and fluorous synthesis. Top. Curr. Chem. 206, 79–105. http://dx.doi.org/10.1007/3-540-48664-X_4.

Pangborn, A.B., Giardello, M.A., Grubbs, R.H., Rosen, R.K., Timmers, F.J., 1996. Safe and convenient procedure for solvent purification. Organometallics 15, 1518–1520. http://dx.doi.org/10.1021/om9503712.

Prat, D., Pardigon, O., Flemming, H.-W., Letestu, S., Ducandas, V., Isnard, P., Guntrum, E., Senac, T., Ruisseau, S., Cruciani, P., Hosek, P., 2013. Sanofi's solvent selection guide: a step toward more sustainable processes. Org. Proc. Res. Dev. 17, 1517–1525. http://dx.doi.org/10.1021/op4002565.

Reichardt, C., Welton, T., 2010. Solvents and Solvent Effects in Organic Chemistry, fourth ed. Wiley-VCH, Weinheim.

Williams, D.B.G., Lawton, M., 2010. Drying of organic solvents: quantitative evaluation of the efficiency of several desiccants. J. Org. Chem. 75, 8351–8354. http://dx.doi.org/10.1021/jo101589h.

The Research Notebook

8

Chapter Outline

The keeping of a proper research notebook is an essential part of doing any kind of science. The training in this skill that most students receive in the organic teaching laboratory is rarely adequate for the research laboratory setting. This is especially true for industrial chemistry, where the research notebook is in many cases the first documentation of the conception and/or reduction to practice of a chemical idea. The date of conception is not as important to the patenting process in the United States as it once was, but it is still essential that the research notebook adequately describe the idea, that the date be clearly identified, and that the notebook page(s) be witnessed by a scientist qualified to understand the chemical concepts involved. Patent litigation has literally hinged on a particular compound being prepared in two different labs in two different companies on consecutive days.

While the foregoing level of dedication to procedure is unlikely to be observed in the academic setting, instituting good habits in the keeping of a research notebook while in training will make the transition to the industrial setting that much easier for the chemist. Chemists who are attempting to repeat reactions done by earlier members of a research group will rapidly come to appreciate their predecessors who provided detailed procedures, and to scorn those who provided sparse details of how they actually got that key compound purified or obtained that excellent yield. The experience of trying to replicate an experiment from another person's notebook is excellent training for any novice chemist. It also likely provides an understanding and appreciation for the necessity to adequately document one's reactions in the notebook.

8.1 Paper Notebooks

For synthetic reactions, each research notebook page will generally have a specified set of elements. Each new reaction should begin on a new page. A structural equation describing the reaction or a structure of the compound under study should be at the top of the page for easy visual retrieval. Preparation of a *complete* table of reagents and products (with molecular weights, masses, moles, and mole ratio for each, plus

Handbook of Synthetic Organic Chemistry. http://dx.doi.org/10.1016/B978-0-12-809504-1.00008-X

other relevant physical properties such as density or boiling point) is a lesson many students have learned in the teaching lab. Too often it is the only element consistently found on notebook pages. This table not only helps ensure that reagents are used in the proportions intended, but it also facilitates calculations of the theoretical yield or the volumes of reagents to use. If a literature reference or references are being followed in the execution of this reaction, citations should be provided. If the apparatus used for a reaction is at all exotic or unusual, drawings or pictures should be provided. Considering the pervasiveness of digital cameras, the latter may become much more common. A narrative of how the reaction was actually conducted should be handwritten in the notebook while the experiment is being performed. Any specific experimental observations (color changes, exotherms, bubbling, etc.) should also be provided.

For characterizing the outcome of reactions, there are typically several different types of data that are not amenable to inscription in the notebook, including spectral or chromatographic plots. For each spectrum or piece of data that is collected but not placed directly in the notebook, a unique identifying code [e.g., a notebook identifier (often the chemist's initials along with the notebook number), the page number, and the spectrum number] should be written in the book, along with a very brief indication of what the data show. The same identifier should be placed on the spectra, plots, and other exhibits. These data can be collected in a companion binder that carries an identifier that links it uniquely to that specific notebook.

To make it as easy as possible to locate specific experiments in each notebook, the first 10 to 15 pages should be left blank. This area can be used to maintain a table of contents. Like the reaction equation at the top of each page, table entries should be structural equations for easy visual retrieval. When a reaction is repeated, a page number should be added to the equation that already exists in the table. This way it is simple to locate all of the trials of a particular reaction over time and observe the ways in which it was conducted. Finally, the table of contents must be kept up to date.

An example of one page of a research notebook prepared by an early graduate student in the author's lab is provided (Fig. 8.1). It is a model of how notebooks should be kept. Alas, the author cannot take credit for training this chemist in keeping a notebook, because before beginning graduate school he had already worked in industry, where their much higher standards had already been instilled in him. The reaction that he was conducting is more easily seen in Fig. 8.2. Note that he dated the page and referred to an earlier page in the notebook with another preparation of this compound. He provided the source (in this case, an earlier reaction) of his starting acid. He had an available solution of diazomethane of known concentration (though its source is not mentioned—a minor omission). He drew a picture of the TLC of the reaction mixture, specifying the eluting solvent and the stain used. Potassium permanganate was useful in this case because the reactant and product include an alkene. He recorded a starting time for the reaction and noted the gas evolution (expected in the conversion of an acid to an acid chloride with oxalyl chloride).

When he isolated the product of the first step, he obtained an IR spectrum to demonstrate conversion of the acid to the acid chloride. For this reaction these data are far more informative than TLC, as the acid chloride would simply hydrolyze on the silica gel, returning the acid and suggesting that no reaction occurred. He described that he added the acid chloride to the diazomethane. (It is necessary that diazomethane be in

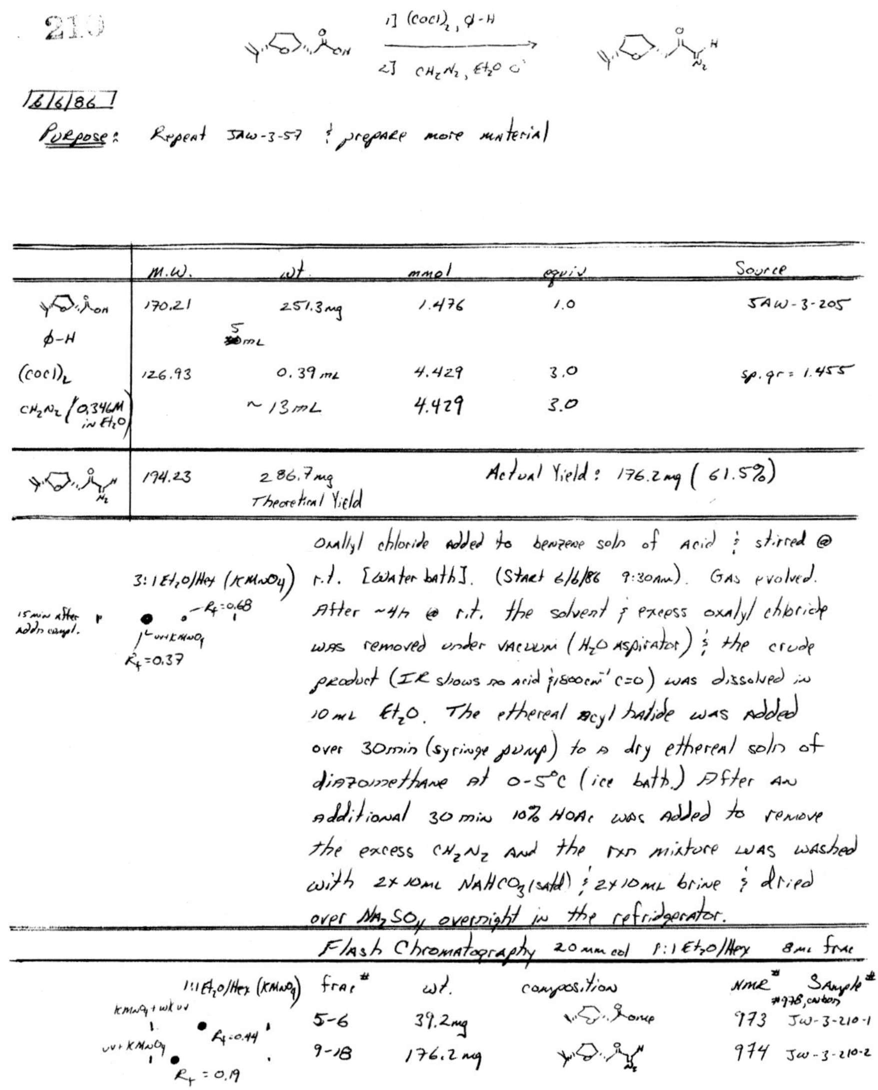

Figure 8.1 A page from the research notebook of John A. Werner.

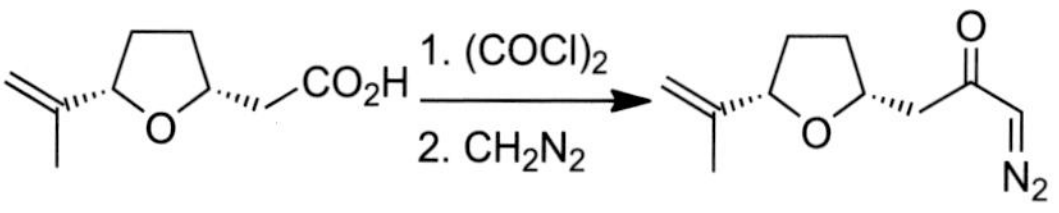

Figure 8.2 The reaction conducted by Werner.

excess to obtain the desired α-diazoketone, and as described in Chapter 5, Section 5.5 this can be accomplished by slow addition.) He described how addition was performed and over what time. He used acetic acid to quench the reaction, which reacts with any remaining diazomethane to form methyl acetate, which can be easily removed by evaporation. He described the phase partitioning workup that would remove any free acids (starting material or acetic acid), and he dried the reaction mixture with sodium sulfate (overnight). The diazoketone might be sensitive to acidic drying agents such as magnesium sulfate. He purified the crude product by flash chromatography and specified the eluting solvent, the size of the column, and the size of the fractions. Note that his solvent for this preparative column is less polar than used for his analytical TLC. He identified a side product in a less-polar fraction, the methyl ester of the starting acid, by NMR. He determined the mass of each of his fractions, and calculated an overall percent yield.

8.2   Electronic Notebooks

With the current trend for all aspects of society to become digitized, there has been a similar push to replace the time-tested paper laboratory notebook with an electronic version. Electronic laboratory notebooks (ELNs) have been available for a broad range of scientific fields for some time, primarily in large organizations. Much early software for ELNs was developed on the "enterprise" model, with a network of users of accessing it on a server. The capabilities of many such programs were evaluated by Rubacha et al. (2011), but these tools were neither tailored to chemical research nor readily available at the smaller scale of academic chemistry departments or research groups. That study is also dated in the fast-changing world of high technology. A 2012 pilot study of enterprise ELN software by one research university found that chemists were not well served by the product(s) considered. Most ELN products available in 2016 are still based on the enterprise software model that is difficult for individual labs to adopt or are not designed with synthetic chemistry in mind.

An ELN tool that is well suited to synthetic chemistry is a package called Elements. Its name does not refer to the chemical elements but to the many different electronic objects that can be constituted onto the window that captures each experiment. It is browser-based and accessed by subscription. It enables users to create multiple notebooks and multiple experiments within each notebook. A sample experiment page is provided in Fig. 8.3.

The top window was created using the element ChemDraw, which provides a blank drawing window that includes a table of reagents and products. This element has functionality that is familiar from the stand-alone ChemDraw application software, but it is not identical to it. Compounds can be added into this window by chemical name if they are common enough to be in Elements' database (e.g., diisopropylamine). If so, a button selects if they will be reactant, product, or reagent (over the arrow). Otherwise, compounds can simply be drawn in place. Cutting and pasting structures or whole reactions out of the regular ChemDraw software is also possible. Each structure is automatically recognized, given a Roman numeral, and entered into the table. Its name, formula, and molecular weight are automatically generated, and any information about it that is in the database, like density, also appears. Key values that are not in the database, like density of less common reagents or the concentration of reagent solutions, can be entered in the table. The chemist also enters the mass or volume for

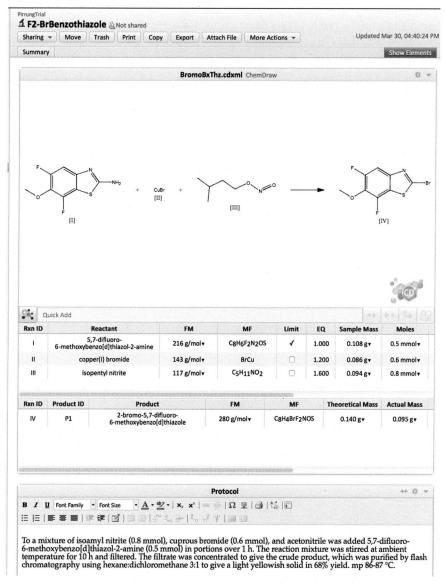

Figure 8.3 An experiment page in the Elements electronic laboratory notebook.

each reagent, and Elements calculates moles and theoretical yield and identifies the limiting reagent(s).

The bottom window was created using the element Protocol, a simple text editor that can be used to enter the written procedure. By adding another element, File, ancillary data for the reaction in any form (spreadsheets, PDF files, image files) can be added to the experiment. It can be used to add NMR spectra, for example. All manipulations of the experiment window are captured in real time, so it is unnecessary to save it or fear loss of work by crashes. This feature is related to the needs of some users to have a validated time/date for experiments owing to intellectual property or other issues. A complete experiment can be exported to an archive or converted to a printable form. Early experience using this tool shows that it has the functionality to replicate the paper notebook and provides some enhanced capabilities. It alleviates the burden of and potential for error in stoichiometry calculations. However, the electronic format may prod the chemist to spend extra time on attractive drawings and more proselike experimental descriptions, rather than the punchy and to the point descriptions of many paper notebooks.

Reference

Rubacha, M., Rattan, A.K., Hosselet, S.C., 2011. A review of electronic laboratory notebooks available in the market today. J. Lab. Automat. 16, 90–98. http://dx.doi.org/10.1016/j. jala.2009.01.002.

Conducting the Reaction Itself

9

Chapter Outline

9.1   Reagents Supplied as Dispersions

Some reagents that would normally be air sensitive are rendered less so by being dispersed in mineral oil. Examples are LiH, KH, NaH, and Li. The mineral oil can cause problems in the workup (chromatography is required unless the product can be distilled), so it is a good idea to remove it before the reaction. This is done as follows: the necessary amount of dispersion (figure by the weight%, usually valid to within 5%) is placed in a dry flask under nitrogen containing a stir bar. The dispersion is covered with (suspended in) dry pentane or petroleum ether. The suspension is stirred for a minute and the stirrer is stopped. The reagent is allowed to settle, and the supernatant is removed by syringe or pipette. This procedure is repeated several times. Tipping the

Handbook of Synthetic Organic Chemistry. http://dx.doi.org/10.1016/B978-0-12-809504-1.00009-1

flask and letting the solvent flow off the reagent (which will clump and stick to the flask) helps remove all of the supernatant.

9.2   Azeotropic Drying

Some reagents, such as nucleosides, are so hygroscopic and hold onto water so tightly that it may be impossible to adequately dry them before they are placed in the reaction flask. In such cases in situ drying using an azeotrope may be effective. The reactant is dissolved in a solvent that forms a minimum-boiling or positive azeotrope with water. Examples are collected in Table 9.1. While many chemists are most familiar with the benzene–water azeotrope, several other azeotropes use far less-toxic organic solvents and contain a greater proportion of water. Of course, the greater the fraction of water, the less the boiling point of the azeotrope is lowered compared to water, so trade-offs must be made. The solvent is evaporated on the rotary evaporator, provided that it is set up such that the vacuum can be released by refilling the evaporator with an anhydrous atmosphere. If this capability is not available, the solvent can be evaporated on a vacuum line with an inert atmosphere refill capability. The solvent is stirred magnetically while vacuum is applied and the flask is held in a water bath that provides the heat of vaporization. A cold trap condenses the evaporated solvent. To completely dry a reagent, evaporation from an azeotropic solvent several times may be required.

Other nonaqueous azeotropes that might be utilized to remove poorly volatile solvents include acetic acid/heptane (boiling point 92.3°C, 70 wt% heptane, 55 mol% heptane) and ethylene glycol/toluene (boiling point 110.2°C, 93.5 wt% toluene, 91 mol % toluene).

9.3   Stoichiometry

An important decision to be made by the chemist concerns the exact amount of each reagent to use in a reaction. Consider the kinetic aldol condensation between ethyl isopropyl ketone and benzaldehyde (Eq. 9.1). The enolate is generated by adding the

Table 9.1 **Low-Boiling Water–Organic Azeotropes**

Azeotrope	Boiling Point (°C)	Water (wt%)	Water (mol%)
Water–pyridine	92.6	43.0	75.0
Water–toluene	85.0	20.2	52.3
Water–heptane	79.2	12.9	45.1
Water–acetonitrile	76.5	16.3	31.0
Water–ethyl acetate	70.5	8.1	29.9
Water–benzene	69.4	8.9	29.8
Water–cyclohexane	69.8	8.5	30.9
Water–hexane	61.6	5.6	21.1

ketone to lithium diisopropylamide (LDA), itself generated by treating diisopropyl-amine with a solution of *n*-butyllithium in hexanes. Benzaldehyde is added and the reaction is quenched with aqueous ammonium chloride. After aqueous workup, the syn aldol product is obtained. A balanced equation shows that each of the organic reagents is required in equimolar amounts, yet most chemists would not conduct the reaction that way. They would be more likely to use the ketone as the limiting reagent, 1.1 molar equivalents of *n*-butyllithium, 1.15 molar equivalents of diisopro-pylamine, and 1.0 molar equivalents of benzaldehyde. The rationale behind these choices follows.

$$(9.1)$$

To obtain a single, kinetically defined enolate geometry, it is important to have LDA in excess during enolate formation. If any excess ketone is present it could serve as an acid to protonate the enolate. While this looks like a nonreaction, such proton transfers would in fact serve to equilibrate the enolate. That is, the enolate formation will no longer be under kinetic control, and the enolate stereochemistry will be defined by thermodynamic stability, which might not favor the single (Z)-enolate needed for the stereoselective aldol reaction. Since the ultimate base in this reaction is the *n*-butyllithium, it must be used in (a calculated) excess compared to the ketone. Use of a 10% excess allows for the actual titer of the *n*-butyllithium to be a bit low. This might be due to inadvertent exposure of the reagent solution to mois-ture since its last titration. It is also important to ensure all of the *n*-butyllith-ium is consumed in generation of the LDA. If *n*-butyllithium remains after LDA formation, which might be due to an inaccurate (too high) titer or a fault in the measurement or delivery of diisopropylamine, it can add nucleophilically to the ketone or aldehyde. That is the reason for using an excess of diisopropylamine compared to the *n*-butyllithium. Because diisopropylamine is basic (removable by an acid wash during extraction), water soluble, and volatile, use of excess diiso-propylamine creates no difficulty in purifying the reaction product. Thus, choices for the amounts of reagents to use in this reaction trace directly to experimental uncertainties concerning their titer or the quantities actually delivered. Finally, while these are the target stoichiometries, the actual molar quantities used will be affected by the precision with which volumes can be measured by syringe (see later). In Eq. (9.1) are given the actual volumes of each reagent that would be used. The slight variation of each molar quantity from the target is due to the practical limitations on measuring and delivering the reagents.

The preceding example demonstrates how valuable the density of a reagent can be to conducting a reaction, as it can simply be measured by syringe. If the density of a liquid is not available from literature sources, it can be determined as follows: Fill a 1 mL tuberculin syringe (with 0.01 mL gradations) and needle (see the techniques following) with the liquid and weigh it. Expel the liquid until the syringe reaches its stop, but take no other action to expel the liquid in the needle. Because a syringe is a TD (*to deliver*) volumetric device, it is made to deliver a spec-ified volume, so only by stopping at this stage can the volume that was expelled from the syringe be known. Weigh the syringe again, and determine the weight of the contained liquid by difference. This method should be accurate to within a few percent.

Reactions involving basic reagents and the generation of conjugate bases from reactants are quite common. Consequently, knowledge of the acidities of a wide range of organic functional groups can be essential to understanding and properly conducting synthetic reactions. It should also be noted that the acidities of com-pounds might be quite different in aqueous media and in organic solvents. It is impossible to directly determine the acidity of a compound that is less acidic than water in an aqueous medium (as its conjugate base simply deprotonates water). Acidities have been an active area of research in physical organic chemistry for decades. Two compilations of functional group acidities, one relative to aqueous media and one in DMSO, are provided in Appendices 8 and 9. These are represen-tative and general data for the functional groups presented, and can be modified by knowledge of the effects of specific structural variations on acidity (i.e., inductive, conjugation, and geometric/hybridization effects). The acidities of specific com-pounds are also available from several compilations. These are frequently available on the Internet and can be easily located by a search engine, one of the best known being the Evans pKa table.

The acid–base and chromophoric properties of certain organic compounds can be exploited to establish with certainty that a deprotonation reaction is complete. For example, triphenylmethide anions are intensely red in many ethereal solvents (an exception being lithium triphenylmethide in diethyl ether). Triphenylmethane can therefore be used as an indicator that base (it has been used with metal amides like LDA or the anion of DMSO) is present in excess in the deprotonation of any compound less acidic than itself (Eq. 9.2). Again, reference to acidity compila-tions of organic functional groups in Appendices 8 and 9 or specific compound acidities will show which anions can be detected in this way. Another indicator class is not dependent on its deprotonation. 2,2′-Dipyridyl or 1,10-phenanthroline give a purple-red color in the presence of lithium dialkylamides in ethereal solvent. Therefore inclusion of these indicators in a reaction can demonstrate a reagent (X–H) has been fully converted to its conjugate base (Eq. 9.3). These indicators also show the presence of alkyl lithiums, as described in titrations (Chapter 3, Section 3.4).

$$(9.2)$$

$$(9.3)$$

9.4   Syringe and Inert Atmosphere Techniques

There are several basic types of glass syringes. Tuberculin syringes with metal Luer lock fittings (Fig. 9.1) are preferable for working with air-sensitive compounds. The Luer lock anchors the needle onto the syringe by screwing it into the syringe. Multifit syringes (Fig. 9.2) usually have Luer lock fittings also, but they tend to leak between the barrel and the plunger. Tuberculin syringes with glass tips (Fig. 9.3) hold the needle on the syringe only by friction. This risks it coming off at an inopportune time. Glass syringes range in size from 100 to 0.5 mL. There is a greater tendency to leak around the plunger in the larger sizes (>20 mL), so they should be avoided when transferring hazardous compounds. During the transfer of such compounds (i.e., neat Me$_3$Al), a thin

Figure 9.1 Glass syringe with Luer lock fitting.
Copyright Cadence, Inc.

Figure 9.2 Multifit syringe with Luer lock fitting.
Copyright Cadence, Inc.

Figure 9.3 Glass tuberculin syringe with ground glass tip.
Copyright Cadence, Inc.

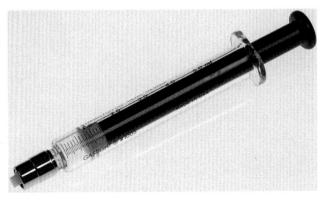

Figure 9.4 A gas-tight syringe with Teflon plunger.
Courtesy of Hamilton Co., Reno, Nevada.

film of grease at the end of the plunger may provide a better seal and help to prevent clogging. Gas-tight syringes with Teflon plungers (Fig. 9.4) are another major class. These are available in a wide range of sizes, from microliters to many milliliters. They are highly recommended when handling dangerous, pyrophoric, or highly air-sensitive reagents. They are available with Mininert valves that prevent the contents from being inadvertently expelled.

Plastic, disposable syringes with rubber plunger seals are also available. Testing of the integrity of these syringes to a given organic reagent or solvent is recommended prior to an actual transfer. Mostly, these are used to measure and transfer aqueous solutions; many labs would not recommend their use for sensitive or dangerous reagents. Their use for transferring gases is described in Section 9.7.4.

Typical stainless steel needles are 20 gauge (20 ga) and come in lengths ranging from 3 to 45 cm. They are flexible and allow fast flow of nonviscous solvents. When clogging or slow flow is a problem (i.e., with organometallics or viscous solvents), a larger diameter needle (18 ga) is used. Care must be taken when using the larger needles because they are easily kinked, and very gentle bends are required. Needles larger than 18 gauge have thicker walls and so are not at all flexible.

For the transfer of substantial volumes, cannulation may be preferable to the use of a large syringe (Fig. 9.5). A cannula is a needle with two points, or two needle points connected by a flexible length of Teflon tubing, which some vendors offer under the name transfer lines. They are used to transfer solution from one vessel directly into another, and they can even be used with a reaction mixture containing fairly reactive intermediates.

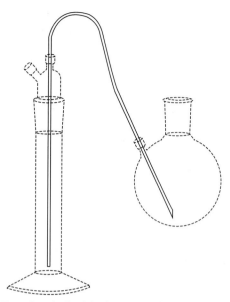

Figure 9.5 Transfer of a liquid using a cannula or double-ended needle.

9.5 General Procedure for Transfer of Materials by Syringe

Very light greasing of the needle is sometimes performed to allow it to puncture rather than tear the septum. The reagent bottle, flask, or reservoir may be attached to an inert gas source by a needle. This may be an inert gas line connected to a bubbler or a balloon filled with nitrogen. If the bottle is *not* attached to a gas source, the syringe should first be flushed (by repeated filling and expulsion) with inert gas and finally filled with gas to the desired volume of the transfer. This gas is then injected into the reagent bottle and the rest of the procedure is followed from step 2.

Step 1—Insert the needle through the septum but not into the liquid. Withdraw some gas into the syringe and remove the syringe from the bottle, grasping the needle as the syringe is pulled out. Expel the gas and repeat.

Step 2—Insert the needle below the surface of the liquid and withdraw more than the desired amount of liquid. Grip both the needle and the plunger. Gas pressure can push the plunger out of the syringe barrel, and needles not secured to the syringe by a Luer lock can fall off easily. There will almost always be a gas bubble in the syringe, so be sure to take steps to eliminate it. Invert the syringe, bending the needle into a U. The gas bubble should now be at the tip of the syringe. Expel gas and liquid until the syringe reads the desired volume. Pull the plunger back, withdrawing a bubble of inert gas into the syringe. This protects the liquid from exposure to air during transfer and prevents dripping. Grasp the needle and remove it from the bottle.

Step 3—Insert the needle through the septum of the receiver and hold the syringe with the tip up (a long needle may be needed to do this). Push the plunger in until it stops, injecting the small bubble of gas and the liquid that were in the syringe (drop-wise if needed). The needle will remain filled with liquid. Pull back the plunger and remove some nitrogen into the syringe. Withdraw the needle and syringe from the receiver.

If a corrosive or chemically reactive reagent was transferred by a syringe, it is important to clean the syringe and needle by flushing them with a solvent immediately following the transfer. This prevents them from becoming clogged or frozen. For lithium reagents, hexanes can be used, which dilutes the reagent. The hexane washes should be carefully added to ethyl acetate and the syringe and needle flushed with the hexanes/ethyl acetate mixture several times. For Grignard reagents, ether should be used in place of hexanes, and the rest of the procedure followed as described. For other reagents, acetone can be used.

For the measurement and transfer of volumes smaller than 0.25 mL (250 μL), microliter syringes are used. These are usually available in 100, 50, and 10 μL sizes. Techniques are the same as above except that the needles are short and not flexible, and most often integrated into the syringe. The containers usually must be inverted to immerse the tip of the needle in the liquid and to expel all gas from the syringe. Alternatively, the plunger can be pumped several times with the needle in the liquid. These syringes sometimes have Teflon plungers, which must never be placed in the drying oven. They can be dried by placing disassembled syringes in a warm spot, like on top of an oven (a useful source of the c. 60°C used to dry NMR tubes).

The pervasive use of syringes to add solutions of reagents and the convenience in measuring reagents based on molar concentration has induced chemists to use the same practice with a plethora of neat reagents, for example, diisopropylamine (used earlier in the chapter for generating LDA). The key to this practice is using the density of the compound to calculate the mass and therefore the moles (or, working backward from the moles to the volume). The density is frequently available in the chemical catalog from which the compound was purchased. The main potential snag with this approach relates to a reaction conducted at a temperature below the freezing point of the compound (also often available from the catalog). The neat reagent will turn immediately into a rock upon touching the cold solution, rather than dissolving as hoped.

When performing reactions in an inert atmosphere, special attention should be paid to the integrity of the apparatus. Each piece of equipment should be clamped to maintain its position as manipulations such as septum punctures are performed. Each stopcock, stopper, and tubing connection should be wired on to prevent breaches in the case of pressure surges.

9.6 Addition

If the order in which reagents are added has not been defined in a precedent for a reaction, it is crucial to consider this question carefully. Often a reagent is already present in the reaction mixture and another reagent is added to it because it might be difficult to transfer, or simply because this is easier. A classical example of this situation is in Grignard reactions, where the organometallic reagent is first generated in an ether solution and then the electrophile is added. This is the so-called normal addition procedure. An "inverse" addition involves addition of the Grignard to the electrophile. Transferring the Grignard naturally increases the risk of its exposure to the atmosphere and is not preferred. However, consider the addition of methyl Grignard to ethyl levulinate. In general, ketones are more reactive to nucleophilic addition than esters, so the chemist might expect to be able to selectively add to the ketone to give the hydroxyester product (Eq. 9.4). Yet, what would happen if ethyl levulinate were added to a methyl Grignard solution? When the first drop of ethyl levulinate hits the solution, undoubtedly the ketone would react with the Grignard first, but a large excess of Grignard reagent would still be present. It would certainly react with the ester as well, and the reaction outcome would not be the hoped-for selective addition. Now consider the reverse situation. What happens when the first drop of a methyl Grignard solution is added to the stoichiometric amount of ethyl levulinate? It adds to the ketone more quickly than to the ester, and all of the Grignard is consumed, so the ester remains intact. Only by adding the Grignard to the electrophile can the intended outcome be achieved.

(9.4)

There are at least two schools of thought on procedures for the addition of reagents. One is that solutions should be dropped directly into the reaction mixture so there is no possibility of the reagent freezing, precipitating, or otherwise becoming heterogeneous so it does not mix. Another is that solutions should be added along the wall of the flask so they acquire the temperature of the reaction mixture overall before they mix with other reagents. This latter point applies particularly when the reagent is at room temperature and is being added to a cooled reaction mixture. It is also possible to add precooled reagent solutions. On a large scale, this can be accomplished with a jacketed addition funnel, in which a coolant can surround the reagent solution. On a smaller scale, a double-ended needle (cannula) can be used. It is placed with one end in the solution to be transferred and the other end in the receiver. An inert gas at a higher pressure than in the receiver is used to push the solution through the needle. Whole reaction mixtures may be transferred in this way, even at low temperatures. Cannulation is discussed in more detail in Section 9.7.6.

9.7   Special Techniques

9.7.1   Water Removal

Water is often the product of an organic reaction, for example, in the elimination of an alcohol (Eq. 9.5), a ketalization (Eq. 9.6), or an esterification (Eq. 9.7). The removal of this water will pull the equilibrium forward through Le Chatelier's Principle. A common way to remove it is via a solvent (heptane, toluene) that forms a minimum-boiling azeotrope with water and a Dean–Stark trap (Fig. 9.6). The azeotrope vapor is condensed into the trap. Because the aqueous and organic components are immiscible, two liquid layers form in the trap, and the lighter one (the organic) eventually overflows back into the pot. The trap has volumetric markings that enable the volume of water to be measured (and in some cases a stopcock that enables the water to be removed). Reaction progress can be determined by the volume of water produced. In practice, it is often useful to wrap this whole assembly in aluminum foil to retain heat. It may otherwise be difficult to induce solvent vapors to reach the condenser. Another trick sometimes used is to add a small amount of water to the trap initially to ensure that layers are formed.

$$(9.5)$$

$$(9.6)$$

$$(9.7)$$

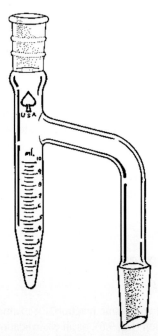

Figure 9.6 A Dean–Stark trap collects in the blind reservoir the water produced from a reaction. Volume graduations on this reservoir enable reaction progress to be followed by the amount of water that has been removed. In some traps the reservoir can be drained by a stopcock.

(A) **(B)**

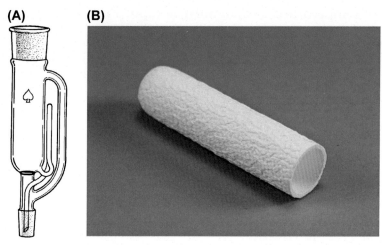

Figure 9.7 (A) The Soxhlet extractor, which is used with a thimble. (B) This thimble is made from rigid paper and is available in a range of sizes to fit different-sized extractors. The thimble contains the solid, preventing it from being carried by the liquid to the siphon tube or into the flask. Fritted glass thimbles are also available.

An alternative to the Dean–Stark trap uses a Soxhlet extractor (Fig. 9.7). Like the Dean–Stark trap, this apparatus benefits from being wrapped in aluminum foil to prevent radiative heat loss. The Soxhlet extractor repeatedly suspends a solid in a liquid to move a substance from one phase to the other. The solvent is heated to boiling, and the vapor moves up and is condensed into the upper reservoir. This reservoir contains the solid in a paper thimble, which is where phase transfer occurs. Once the reservoir is filled, the liquid overflows a siphon tube and the reservoir is drained into the flask below. A solute can be transferred from the solution to the solid, as intended for the removal of water from a reaction mixture using activated molecular sieves in the thimble. Another example uses a Soxhlet extractor to drive the exchange of carboxylic acids on a metal complex (Eq. 9.8). The acetic acid forms an azeotrope with chlorobenzene and, upon reaching the thimble filled with sodium carbonate and sand, is irreversibly converted to sodium acetate. A substance can also be transferred from the solid to the solution, as when commercial copper iodide is extracted with tetrahydrofuran, which removes CuI_2 from CuI because CuI_2 is much more soluble in tetrahydrofuran. This is an excellent method to purify CuI for the generation of organocuprates.

$$Rh_2(OAc)_4 + RCO_2H \rightleftharpoons Rh_2(O_2CR)_4 + AcOH \xrightarrow[Na_2CO_3]{PhCl} NaOAc \tag{9.8}$$

9.7.2 Reactions Above Atmospheric Pressure

Some reactions must be conducted in closed systems at pressures above atmospheric, for example, when the reaction temperature is well above the boiling point of the

Table 9.2 **Temperatures (°C) of Solvent Vapor Pressures**

Solvent	5 atm	10 atm	Solvent	5 atm	10 atm
Ethyl ether	90	122	Ethyl acetate	137	170
Pentane	92	125	Cyclohexane	146	184
Methanol	112	138	Fluorobenzene	148	184
Acetone	113	144	Heptane	166	203
Chloroform	120	152	Toluene	178	216
Ethanol	126	152	Acetic acid	180	214
Isopropanol	130	156	Chlorobenzene	205	245
Hexane	132	167			

Data from CRC Handbook of Chemistry and Physics, sixty second ed., p. D-189-190.

solvent or a reactant. Gaseous reagents (H_2, CO, NH_3) may also permit or require elevated pressure. In designing experiments under these conditions, it is important to understand how heating affects pressure and the pressures reaction vessels can tolerate. Coyne's view (2005) is that standard 1/2 in. glass tubing can withstand 150 psi internal pressure, and heavy-walled tubing can withstand 400 psi. The main influence on pressure in a sealed system will be the vapor pressure of the solvent at the reaction temperature. That should not be considered the last word on the subject, though, because reactions that generate gaseous products will experience major pressure increases. Temperatures are listed in Table 9.2 at which some commonly used solvents have vapor pressures that may be tolerated by some reaction vessels.

Safety Note

Further consultation with a glassblower on these methods may be worthwhile. Any pressurized system is hazardous and should be protected by a safety shield. Specific and complete safety information concerning the pressure vessels or tubes being used should be sought before any such reactions are attempted. Any defects (chips, scratches, bubbles, cracks) in any tubes or vessels covered in this section can lead to catastrophic failure and/or explosions. In an abundance of caution, some workers use only new reaction vessels or tubes for reactions at pressure and retire them from service after a single use.

If a solvent or pressure of interest is not listed, a method to estimate solvent vapor pressure at various temperatures is based on the Antoine equation (Dreisbach and Spencer, 1949). It uses the near-linear relationship seen in a plot of the log of vapor pressure (P in Torr) versus an inverse temperature term (T in °C). This equation also has two empirical terms A and B (Eq. 9.10). For solvents commonly used in organic synthesis, with boiling points in the 50–100°C range, A is ~7, enabling the equation to be solved for B based on the solvent's boiling point at atmospheric pressure (Eq. 9.10). The vapor pressure at a new temperature can then be estimated using Eq. (9.11). This

Table 9.3 Conversions Among Four Common Pressure Units[a]

Term (Abbreviation)	Bar	Atmosphere	Torr	Pounds per Square Inch
Bar (bar)	1	0.987	750.06	14.50
Atmosphere (atm)	1.013	1	760.00	14.695
Torr (Torr)	1.333×10^{-3}	1.315×10^{-3}	1	1.933×10^{-2}
Pounds per square inch (psi)	6.89×10^{-2}	6.70×10^{-2}	51.714	1

[a]A single unit in the left-most column corresponds to the indicated number of units in columns 2–5.

value should approximate pressures a sealed reaction vessel would experience with the solvent at that reaction temperature.

$$\log_{10} P = A - \frac{B}{T + 230} \tag{9.9}$$

$$B = 4.119 \times (bp + 230) \tag{9.10}$$

$$P = \text{antilog}_{10} \left(7 - \frac{B}{T + 230}\right) \tag{9.11}$$

In literature concerning reactions at pressure, terms and units can be confusing because usage has changed over time. The term mm Hg is archaic but has nearly the same value as Torr. The distinction between psi and psig is that the latter is read from a pressure gauge, meaning it is the pressure in psi *above* atmospheric. The SI unit for pressure is Pascal (abbreviated Pa), which is rarely used in synthetic chemistry; it is equivalent to 10^{-5} bar. Table 9.3 provides conversions among pressure units in current usage. Internet tools can be found to perform such conversions: http://easyunitconverter.com/pressure-unit-conversion/pressure-unit-converter.aspx.

Reactions at elevated pressure can be challenging to perform on the smaller scales covered in this book. One approach to this problem is the use of a sealed tube (Fig. 9.8). A glass rod is welded to the top of the tube using a flame. Reactants are added and the bottom of the tube is cooled in a dry ice bath or liquid nitrogen. The constriction in the tube is heated while the glass rod is pulled to draw out the constriction to closure and separation. Following the reaction, the bottom of the tube is again cooled (slowly and carefully! fracture is possible) in a dry ice bath or liquid nitrogen before the tube is opened, either by scoring and breakage or by a flame.

Heavy-walled pyrex tubes with a stainless steel lid (Fischer–Porter tubes, often available from glassblowers) can tolerate pressures up to 20 atm, make the sealing and opening operations much less trouble, and can include fittings and gauges to assist in preparing and monitoring reactions. Resealable pressure tubes (Fig. 9.9) are easier to use and are rated to 150 psi at 120°C.

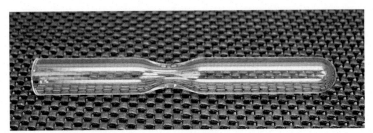

Figure 9.8 Pyrex pressure tube is sealed after addition of reagents by drawing out the constriction.

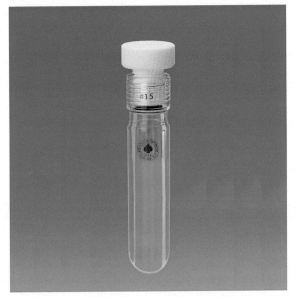

Figure 9.9 Resealable pressure tubes have a threaded Teflon cap that screws on to seal the tube.

Another option for a resealable pressure tube is the Q-tube (Fig 9.10). These heavy-walled borosilicate glass tubes are rated to pressures of 500 psi at 180°C. The pressure monitoring seal safely releases gas at a set point (120 or 180 psi), then reseals once the pressure drops. Gauges and other fittings are also available for Q-tubes.

Some laboratory microwave reactors (Chapter 6, Section 6.1.2) are set up to monitor pressure as well as handle pressures significantly above atmospheric, and these may be a more convenient alternative to the sealed tube. Microwave digestion bombs made from Teflon (Fig. 9.11) are another. These tolerate working temperatures up to 250°C and pressures up to 1200 psi and are sealed merely by turning a threaded cap. They accept reaction volumes of 23–45 mL and are designed for use in microwave ovens, not the specialized microwave synthesis reactors that were discussed earlier.

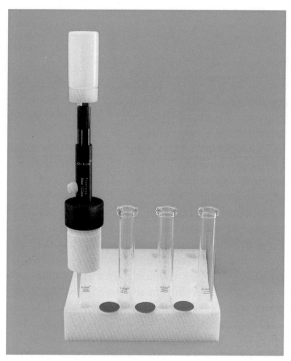

Figure 9.10 The Q-tube is loaded with reagents (up to 1.2 g of starting materials per 12 mL tube) and solvent (to one-third its volume) and sealed to the pressure monitor with a blue Teflon septum.

Figure 9.11 Teflon bombs for microwave heating under pressure. Photograph provided by Parr Instrument Company, Moline, Illinois.

9.7.3   Reagent Gases at Elevated Pressure

A reaction that requires a reagent gas to be used above atmospheric pressure adds to the challenges discussed earlier. Examples of such reagents include carbon monoxide and hydrogen. Specialized equipment is available to conduct carbonylations and hydrogenations at elevated pressures (such as an autoclave) but is also designed to work primarily on a large scale. The Parr shaker (Fig. 9.12) is useful for catalytic hydrogenation reactions requiring relatively low pressures, up to 5 atm. These conditions would be effective in reducing highly substituted alkenes and some aromatic compounds. The shaker is rather simple, with a heavy-walled glass pressure bottle sealed by a rubber stopper with a gas inlet. This whole assembly is placed in a metal cage designed to hold the stopper in the bottle and contain the glass should the pressure bottle break. The cage is rocked by an oscillating arm driven by an electric motor, enabling liquid and gas to be mixed vigorously. The reactor has a gas reservoir, a vacuum inlet for evacuate-and-fill procedures, and a diaphragm valve to control the gas pressure. The main drawback of the Parr shaker is the volume of the smallest bottle (250 mL), which is not consistent with the small-scale reactions of exploratory synthetic chemistry. An issue with most reactions of gases at elevated pressure is that the stoichiometry of the gas cannot be controlled. This means reactions cannot be selective for one reducible functional group over another unless one is simply unreactive under those temperature/pressure conditions.

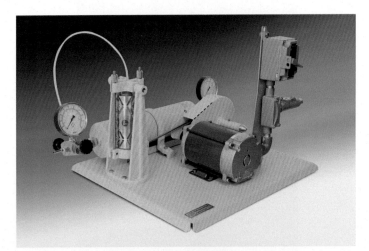

Figure 9.12 A Parr shaker for conducting hydrogenations under moderate pressure with shaking.
Photograph provided by Parr Instrument Company, Moline, Illinois.

Safety Note

Particular care must always be exercised in conducting any reaction involving hydrogen. The catalysts used for hydrogenations typically can also catalyze the reaction of hydrogen with oxygen in air, leading to a fire. Palladium on carbon frequently ignites upon first coming into contact with flammable organic solvents, particularly methanol, and as such represents a significant safety risk. Hydrogenations were a leading cause of laboratory fires in a recent informal survey of US academic chemistry departments.

These hazards can be reduced by beginning and ending every catalytic reaction of hydrogen with an inert atmosphere. The catalyst is added to the reaction vessel and it is placed under an inert gas. The solvent is added and the gas is switched to hydrogen by an evacuate-and-fill procedure. After the reaction is complete, the atmosphere is switched back to an inert gas before the reactor is opened. The goal is to expose the catalyst only to inert gas and hydrogen, never oxygen. Removing supported catalysts is still done by filtration, but even that can be protected from air by the funnel technique of Chapter 5, Section 5.4.

9.7.4   Reagent Gases at Atmospheric Pressure

For reactions requiring a gaseous reagent at only ambient pressure, specialized equipment is not required provided the stoichiometry is not crucial to success (i.e., the gas is the reagent in excess). The usual inert atmosphere in a reaction flask is simply replaced by a pure gas from a tank. An alkene hydrogenation that requires no selectivity over other reducible functional groups is an example of such a transformation.

An alternative uses a latex balloon filled with a reagent gas. The balloon is pulled onto a short length of rubber tubing, filled from a gas tank, and attached to the reaction vessel via a gas adapter. Repeated evacuation and filling of the vessel exchanges its atmosphere for the reagent. When using gases at atmospheric pressure, recognize they do not instantly diffuse to mix with the volumes to which they are exposed; they must be actively moved as described. The balloon may be left connected to the flask during the reaction, providing a pressure slightly above atmospheric (sometimes referred to as "positive pressure") and maintaining that pressure if a significant volume of gas is consumed. Recalling that a mole of ideal gas has a volume of 22.4 L at STP, even relatively small-scale reactions can consume significant volumes of gas (112 mL for a 5 mmol reaction). Examples include carbon monoxide for organometallic transformations and 2-methylpropene (isobutylene) for the formation of *tert*-butyl ethers.

A convenient method to handle reagent gases at atmospheric pressure uses large plastic syringes, such as the 50 or 100 mL sizes. These offer the advantage that the stoichiometry can be controlled through the use of measured volumes. A gas can be transferred into a plastic syringe from a tank, and with the syringe so loaded, it can be moved into a reaction vessel with the plunger. Gases can also be generated inside a syringe (e.g., SO_2

from NaHSO$_3$ and HCl), which is convenient for gases that are used intermittently or that are not available in lecture bottles. However, they are not anhydrous.

A video of techniques for manipulating gases with plastic syringes is available here: https://www.youtube.com/watch?v=VsOAyXY1-7U.

It may also be possible to generate a gas as needed for use in a reaction. For example, ozonolysis of alkenes is commonly performed in this way. Commercial ozone generators are available that use either air or oxygen and a corona discharge to produce ozone in a gas stream. The toxicity of ozone to the human respiratory system demands that such ozone generators be used in an efficient fume hood. The ozone-containing gas stream is slowly bubbled into a solution of the reactant (in methanol, acetic acid, chloroform, hexanes, or ethyl acetate) to conduct ozonation. The rate of ozone production is defined by the input gas composition, its flow rate, and the electrical power, but is around 5 mmol/h with air and 0.5 mol/h with O$_2$. A calibration can be made of the ozone generator by oxidizing a 5% aqueous KI solution mixed with an equal volume of acetic acid and titrating the so-produced I$_2$ with sodium thiosulfate solution to a starch endpoint. This gives the experimenter a good estimate of how many millimoles of ozone can be generated per minute at specific instrument settings and therefore how long a particular ozonolysis reaction should require.

When only one ozone-reactive function is present in the molecule, ozonation to exhaustion is simple to conduct. It is almost like a titration. This approach exploits the fact that solutions of ozone in organic solvents are blue. So long as the ozone is consumed by reaction, the solution will remain colorless. When all of the reactant is consumed, the ozone dissolves, demonstrating the end point with its blue color. Adding a small complication to all this is the fact that ozonolysis of alkenes is generally conducted at −78°C (because ozone reacts explosively with alkenes at room temperature). Therefore these observations must be made through a dry ice cooling bath. Once the end point is reached, the ozone stream can be replaced with a stream of nitrogen and the excess ozone can be swept out of the solution. Workup of the ozonolysis reaction requires reduction of the ozonide at the low reaction temperature, since the ozonide is also a peroxide and such compounds can be dangerously explosive (see Chapter 7, Section 7.2).

Oxygen gas can also be used as a reagent, either in radical reactions as its ground-state triplet or in cycloaddition reactions as its excited singlet state. Simply exposing a reaction mixture to air typically does not promote an efficient reaction with O$_2$. Air must be bubbled through the solution or an oxygen-rich atmosphere must be used. If air is used, moisture and carbon dioxide may need to be removed with a KOH drying tower if the reaction would be sensitive to these species (e.g., oxygenation of an enolate). The generation of singlet oxygen in an oxygenated solution requires light (typically visible light from a sunlamp) and a triplet sensitizer (such as tetraphenylporphyrin or Rose Bengal). The generation of singlet oxygen may be an unwanted side process during photochemical reactions, providing another circumstance in which degassing (deoxygenation) of the solvent is important. Pure O$_2$ gas should be used very carefully because it can promote combustion reactions that are not a concern at the lower partial pressure of O$_2$ in air. Also, since the products of most reactions involving O$_2$ are peroxides, which as a class are reactive and often explosive, such reactions must be conducted with care and appropriate precautions, such as a safety shield.

9.7.5 Ultrasonication

The use of ultrasonic energy may be recommended for some reactions. The ways in which ultrasound can potentially influence reactions have been extensively discussed but will not be belabored here. Reactions for which there is a clear benefit include those involving solid–liquid interfaces or activation of metal surfaces, such the formation of Grignard or organozinc reagents. The effects of ultrasound may include removing oxide coatings from the metal surface, exposing reactive sites of the clean underlying metal, and facilitating reactant transport to and from the solid phase. In fact ultrasonication is believed to be the most effective method of mixing known. The ability of ultrasound to affect a reaction is clearly dependent on the strength of the sonic energy source. Sophisticated ultrasonication instruments may be available in specialized laboratories, but many labs have ultrasonic cleaning baths. The effectiveness that such a bath should have on a chemical reaction can be easily tested. A sheet of aluminum foil is placed in the bath for 30 s. If it emerges pockmarked with holes, the bath is strong enough to affect a reaction.

9.7.6 Pyrophoric Reagents

Compounds that spontaneously inflame upon contact with air are generally defined by chemists as pyrophoric. A familiar example is white phosphorous. There may be other definitions and classifications of particular chemicals as pyrophoric, depending on the jurisdiction and who is making them (occupational safety regulators, fire codes, transportation regulators). Some of the compounds that may be considered pyrophoric include metal hydrides, finely divided metals, and a range of metal alkyls. The pyrophoric reagents most commonly used in the synthesis lab are likely solutions of alkyl lithiums. These are tremendously useful as powerful bases that can generate a wide variety of anions for use as nucleophiles and bases.

Safety Note

Among the essential precautions for use of pyrophoric compounds are a flame-resistant lab coat, an ABC dry chemical fire extinguisher close at hand (not a CO_2 extinguisher), and double gloves, one fire-resistant glove and one chemical-resistant glove.

The most notorious of these alkyl lithiums is *tert*-butyl lithium (*t*-BuLi). It has been traditionally supplied in pentane solution, but heptane solutions are now available. This compound specifically is highly dangerous; extensive training in its use must be received and extensive safety precautions must be exercised before a chemist could even consider using *tert*-butyl lithium. It is so reactive, even a tiny drop of *t*-BuLi solution remaining at the tip of a syringe needle after a transfer can spout an impressive jet of purple flame after removal from the inert atmosphere of the reaction setup.

Accidents with *t*-BuLi have specifically led to injury and death by fire of research workers in basic science laboratories. This absolutely could happen to you. Consequently, *t*-BuLi should be used only when literally no alternatives are available; the regulatory burden to use it will certainly be substantial.

tert-Butyl lithium is one of the most powerful organic bases known, and for that reason is used to make other anions that are difficult to generate. However, it is also so reactive that it can rapidly attack ethereal solvents, even below room temperature ($t_{1/2}$ of 40 min in THF at −20°C). Methods that enhance the basicity of other, safer organolithiums such as *n*-BuLi may enable the same transformation to be accomplished without risking *t*-BuLi. This is typically done by adding lithium complexing agents (HMPA, DABCO, TMEDA) or potassium *tert*-butoxide; the latter converts *n*-BuLi to the potassium alkyl, also known as Schlosser's superbase.

The organolithium most commonly used in the synthesis lab is surely *n*-BuLi. It fits many definitions of pyrophoric, but is safer to handle than *t*-BuLi. It is quite inexpensive on a molar basis because it is an article of commerce; it is used as an initiator of anionic polymerization on an industrial scale, and *n*-BuLi is shipped across country in railroad tank cars. For those who wish to directly generate their own LDA, *n*-BuLi is the base of choice.

Pyrophoric reagents are good examples of the effect of scale on safety risk. Seeing a flare from a few microliters of *t*-BuLi at a syringe tip creates a little excitement but not great concern, whereas mishandling multiple milliliters of *t*-BuLi could trigger a serious conflagration. For transfer of larger volumes of pyrophorics, the cannulation technique described earlier can be used, but must be performed with great care that the cannula not become a hose spewing flaming reagents. Alternatively, multiple small transfers of a solution of pyrophoric with a syringe may be safer. It is especially important to observe the rule that syringe capacity should be twice the volume of liquid to be transferred (i.e., transfer no more than 5 mL in a 10-mL syringe). This means the plunger of the syringe will always remain well inside the barrel, with little risk of separating from it, when the full volume to be transferred has been loaded. It is also essential to use syringes with Luer lock fittings to ensure the needle is not inadvertently detached.

Even when working with pyrophorics on a small scale, some syringe transfer practices must be modified. In particular, conventional glass syringes almost always have some leakage of gas between the plunger and the barrel. If this occurs when the reagent in the syringe is an alkyl lithium, it reacts with moisture in the air to form lithium hydroxide, which precipitates and causes the plunger to stick in the barrel of the syringe. This is really a nightmare scenario for the chemist, having a syringe full of a dangerous reagent and no means of getting it out or getting a quenching reagent in. Therefore, the transfer of pyrophoric reagents should be conducted with a gas-tight syringe (Fig. 9.4). These have a smooth glass barrel and a plunger with a PTFE head that fits tightly in the barrel. There is no leakage of gas between them, and the syringe transfer can be performed with much less risk.

For either method for transfer of a pyrophoric reagent, syringe or cannula, the general techniques described in this book should be reviewed, all institutional safety procedures for pyrophorics should be followed, and then a video on the procedure should be viewed. Because the importance to chemical safety of the proper handling of pyrophorics has become apparent, institutions have recently produced training videos regarding this hazard. A two-video series is recommended:

https://www.youtube.com/watch?v=3_cBVfYVAC8
https://www.youtube.com/watch?v=WUHrzcEunNY

9.8   Unattended Reactions

It is fairly common that a reaction cannot be completed within a working day and therefore is allowed to run unattended overnight. That this can be a relatively safe practice is shown by the many times it has been done without problems. The situations most likely to lead to problems include reactions performed with heating and a water-cooled condenser. If the condenser fails for any reason, the solvent can be lost by evaporation and the reaction mixture taken to dryness. Cooling can be lost if the building cold water service pressure drops, for example. The best outcome in this situation is that merely the compound is destroyed. Many reagents decompose in catastrophic ways when taken to dryness (and heated!). Heated reactions should be monitored and controlled by a thermostat such that heating is discontinued if cooling is lost.

Another reason cooling might be lost, a flood, causes its own safety issues. The condenser is connected by tubing both to a serrated tip on the cold water tap and to the hood drain; if that tubing splits or even just slips off a fitting because the water pressure rises, flooding will continue until it is discovered. Water is quite damaging to lab furnishings and will likely flow to lower levels, damaging other labs and perhaps causing safety issues there as well. Several precautions are necessary to mitigate this hazard. Because the cold water tap is not a precision device, flow from it can fluctuate with service pressure. A needle valve with a hose adapter should be added between the tap and the tubing. This permits the tap to be opened liberally to ensure there is always adequate water flow, but that flow can be controlled precisely with the needle valve to be neither too high or low. Tubing should be inspected frequently and replaced if there is any evidence of damage from chemicals it contacts during use. The tubing should be securely attached at each end to the fittings by wrapping it with copper wire.

If you allow a reaction to proceed unattended for any substantial period of time, it is wise (and required in some settings) to leave in a prominent location, such as on the hood sash in front of the reaction, a short written summary of the experiment as well as the laboratory notebook page for details. All hazards should be specified and a contact telephone number should be provided. All of this information is intended to aid a colleague or first responder who comes upon the reaction in the event of some mishap or an accident.

9.9   Quenching

This term has acquired a variety of meanings and uses, some proper, some not. In principle, it refers to adding a substance that deactivates reagents present in the reaction mixture that can affect product isolation. For example, a Grignard addition to a ketone initially produces a magnesium alkoxide. To isolate the desired alcohol, acid is added to the reaction mixture, which protonates any remaining Grignard as well as the alkoxide (Eq. 9.12). This is appropriately called an acid quench. Loose uses of the term include any solution added following the end of the desired reaction in preparation for the workup.

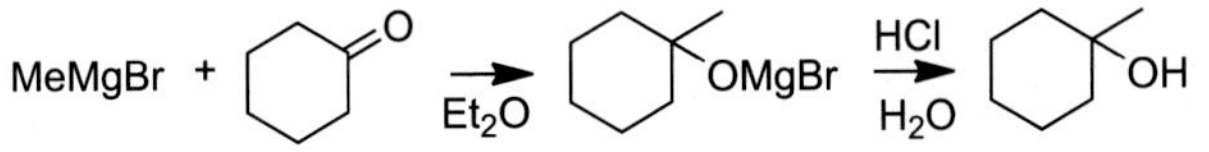

$$\text{MeMgBr} + \underset{\text{Et}_2\text{O}}{\overset{\text{O}}{\longrightarrow}} \quad \text{OMgBr} \quad \underset{\text{H}_2\text{O}}{\overset{\text{HCl}}{\longrightarrow}} \quad \text{OH} \qquad (9.12)$$

9.10   Specialized Reagents

9.10.1   Diazomethane

Diazomethane is attractive as a methylating agent for carboxylic acids and phenols because it reacts quickly and highly efficiently with the production of only N_2 as a by-product (Black, 1983). Its natural yellow color is discharged as it reacts, providing automatic indication of reaction progress. However, because diazomethane is highly toxic, it should be generated and used only in a well-functioning fume hood. Because it explodes on contact with some metals or ground glass of any type (joints, stoppers, syringes, stopcocks), it should be handled behind a safety shield, and other personal protective equipment should be used. Because it has a boiling point of −23°C, it is usually handled in the ethereal solutions in which it is generated. Because it explodes on contact with $CaSO_4$, its solutions or vapors must never be dried with drierite. Despite all of these hazards, it can be worked with safely, provided that appropriate precautions are observed.

Two main methods are used to prepare diazomethane. One uses commercially available apparatus specifically designed for its preparation and distillation while entrained with ether. The resulting ether solution is typically of 0.3–0.4 M concentration and diazomethane is in its purest form. Such apparatus have specialized joints without ground glass and come in a range of sizes for generating diazomethane on scales of around 1, 50, or 300 mmol. The other method uses conventional glassware. Both methods use hydroxide to generate the diazomethane from nitrosamide precursors. The more formal method involves adding *N*-methyl-*N*-nitroso-toluenesulfonamide (Fig. 9.13), also known as Diazald, to KOH. The manufacturer's instructions for the use of this apparatus should be followed explicitly.

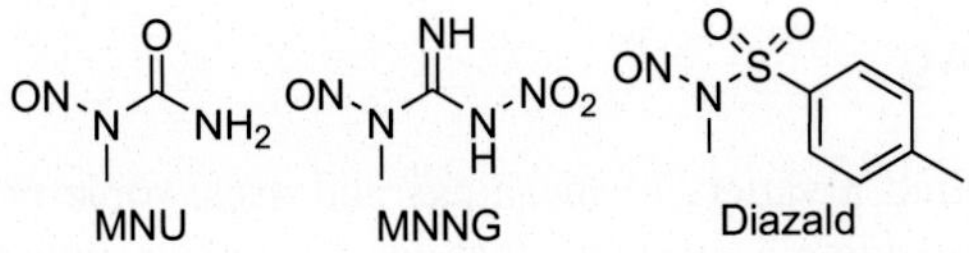

Figure 9.13 Reagents used as precursors to diazomethane.

The home brew method to make diazomethane can be found in its original form in *Organic Syntheses* (using *N*-methyl-*N*-nitrosourea as the precursor) (De Boer and Backer, 1956). This method uses a two-phase system of 50% aqueous KOH and diethyl ether in an Erlenmeyer flask cooled in an ice-water bath and stirred magnetically. The precursor recommended today, because it is safer to store and handle, is the crystalline solid *N*-methyl-*N*-nitroso-nitro-guanidine (MNNG). However, MNNG is still considered toxic, a severe irritant, a carcinogen, and a mutagen, and is typically used for generation of diazomethane quantities of 1 mmol. MNNG is slowly added to the two-phase system portion-wise. Sufficient precursor must be used to allow for materials transfer losses of diazomethane that are inevitable in the incomplete separation procedures described following. A yellow color will develop in the ether phase as the diazomethane is generated. After all of the precursor has been added, the solutions may be stirred for 10 min or so to allow the reaction to complete. The upper ether layer is decanted into a clean flask held in an ice-water bath. DO NOT use a separatory funnel with a ground-glass stopcock to separate the aqueous solution from the ether phase. Another portion of ether is added to the reaction flask, and it is stirred at ice-water bath temperature to extract remaining diazomethane. This ether layer is also decanted into the clean flask in the ice-water bath. This process may be repeated. The combined ether phases are likely to contain some dissolved water, which may be removed by adding KOH pellets and allowing the solution to stand in an ice-water bath for 0.5–3 h. The resulting yellow ethereal solution of diazomethane is ready for use. This procedure can be conducted on up to a 60 mmol scale.

9.10.2   Lithium Aluminum Hydride

LiAlH$_4$ can be briefly handled as the powder in air without much loss of active hydride, but is an excellent reducing agent in ethereal solution. Its solutions are air sensitive, though. Chemists typically think of the reagent as an ionic species (Li$^+$ and the complex ion AlH$_4^-$), which suggests its solubility would be greatest in more polar ethereal solvents such as tetrahydrofuran (THF). However, the solubility of lithium aluminum hydride in ethereal solvents has been determined, and it is twice as soluble in diethyl ether as in THF. It is often considered pyrophoric and should be handled in light of the procedures described in Section 9.7.6, especially in solution.

Safety Note

A recent informal survey of US academic chemistry departments showed LiAlH$_4$ was the most often reported cause of lab fires.

A common transformation using LiAlH$_4$ is the reduction of carboxylic acids to primary alcohols. Because this reaction necessarily proceeds by initial deprotonation of the acid by a hydride from LiAlH$_4$ to give a carboxylate, there may be an inclination

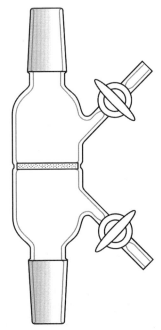

Figure 9.14 Apparatus for filtration of air-reactive solutions under an inert atmosphere.

among chemists to use tetrahydrofuran as the solvent for such an ionic species. It might also be thought the carboxylate would be quite resistant to reduction due to its negative charge and the reaction therefore requires heat, with the higher boiling point of THF providing an advantage. Yet such reductions often proceed as well or better in ether than in THF. A potential reason for this observation is the greater solubility of the reagent in ether.

LiAlH$_4$ ethereal solutions are gray in color, similar to the powder. If they are filtered (under a nitrogen atmosphere, of course, using a Schlenk-ware coarse fritted filter, Fig. 9.14), the filtrate is often much more transparent, and a fine gray residue is left behind on the filter. The residue often exhibits some active hydride, so it must be quenched with care, but the amount of active hydride in the filtrate is typically not appreciably diminished. This filtered ethereal LiAlH$_4$ solution may afford much cleaner reduction reactions. Lore concerning these observations is that Lewis acidic impurities that are insoluble in ether are removed by the filtration process. While evidence about the basis of the effect may be debated, such effects are clearly real in some cases.

9.10.3 Hydrogen Peroxide

This reagent is one of the least expensive and most convenient oxidants known. It is used in a variety of transformations, including the epoxidation of α,β-unsaturated carbonyl compounds and the preparation of peracids. H$_2$O$_2$ is typically available as an aqueous solution in several strengths, including 3%, 30%, and 90%. The first two should be handled with respect, like any chemical reagent, but do not pose undue hazards. Like other peroxides, care must be taken in exposing H_2O_2 to even mildly reducing conditions, which can initiate radical reactions. Chemists should be more concerned about concentrated H_2O_2 solutions (Shanley and Greenspan, 1947), which are commonly used for the in situ generation of peracids such as trifluoroperoxyacetic acid from the anhydride. Contact of hydrogen peroxide of more than 65% concentration with any combustible material (cotton, wood) leads to fire. Pure 90% H_2O_2 does not explode, but matters are different in the presence of a fuel or catalyst. Metal-containing reagents may react violently with H_2O_2. The exothermic decomposition of 90% H_2O_2 yields 5000× its volume in O_2, creating the risk of pressure buildup. The combination of glycerol and 90% H_2O_2 is comparable to nitroglycerin as an explosive, except that the former is 7 times more impact sensitive. Iron, brass, or copper fittings or mercury thermometers should not be used with high concentration H_2O_2 because the metals can promote its decomposition.

9.11   Checklist

The utility of checklists to minimize errors in fields such as aviation, construction, medicine, and investing was admirably advocated in the recent best-selling book *The Checklist Manifesto* by Atul Gawande. Checklists can also be very useful in the synthetic laboratory, and many labs may already be using them. In the spirit of Gawande's book, a checklist for performing everyday synthetic reactions is offered (scaled down) in Fig. 9.15. It will be available in full-size PDF form from this book's website page, and you should enhance its contents for your own research.

9.12   Reaction Time Versus Purification Time

This chapter closes with a broad admonition concerning synthetic reactions. Attention to detail in the execution of the reaction up to this stage will pay enormous dividends in the chemist's time. Conduct of the reaction itself is often a small fraction of the time and effort spent in isolating and purifying the product. Anything the chemist can do to make a reaction more selective and/or eliminate a by-product will be well worth it.

Furthermore, when investigating a pilot synthetic reaction, synthetic chemists of the highest talents isolate and identify *all* products formed, even when they are produced in rather minor amounts. The structures of these compounds can be very informative concerning mechanistic pathways that are diverting material from the desired reaction course. They might also represent chemical transformations that are even more interesting than the reaction the chemist aimed to achieve in the first place, opening doors to new research directions.

REACTION CHECKLIST

☐ Find a procedure to work from: 1. your own notebook; 2. another's notebook; 3. an *Organic Syntheses* prep; 4. a literature experimental.

☐ Consider all safety precautions necessary to use these particular reagents. Plan for proper PPE and waste disposal. Consult lab SOPs for relevant procedures.

☐ Confirm you have all reagents in pure form, or purify them yourself (mostly by distillation). Purify solvents as needed, especially from the solvent system for anhydrous reactions.

☐ Obtain proton NMRs and TLC R_fs for all reagents, and save samples for later comparison.

☐ Start a new page in your notebook with a reaction equation at the top of the page, and enter that page number and equation in the Table of Contents. Record the date and literature cites.

☐ Choose a scale. For many reactions, 1 mmol is a good place to start (if you have that much). If a reagent is commercial and inexpensive (<$5/g) and the procedure looks reliable, consider using multigram quantities of starting material.

☐ Prepare a table of reagents and products in your notebook with all weights/volumes you will use, as well as important related data (like density).

☐ Allow sufficient time to both perform the reaction and complete the work-up. While most crude reaction products can be stored, do not allow reaction mixtures to languish.

☐ Record the procedure you actually execute directly in your notebook as you do it.

☐ Begin the reaction and immediately get a TLC - it could be fast and already complete. Record any other observations during the reaction - bubbling, color, etc. Place a thermometer into most reaction mixtures to monitor the reaction temperature and any exotherms.

☐ Monitor the reaction periodically by TLC. Stop it when all starting material is consumed or no further change is evident.

☐ Work-up the reaction and obtain the crude product. Weigh it to get a mass balance.

☐ Prepare a proton NMR sample and use it to spot a crude TLC. Obtain the NMR, assign known peaks of starting materials by comparison, and mark potential product peaks.

☐ Obtain an IR of the crude product if relevant (i.e., any change in carbonyl groups).

☐ Purify, most often by flash chromatography, but also by recrystallization, distillation, etc.

☐ Obtain proton NMR spectra on each spot/fraction to assign tentative structures. Once you know that a fraction is a pure compound, what it is, and how much you have, calculate yield.

☐ Obtain an IR if the proton NMR is inconclusive regarding structure. Consider getting a MS if the structure is still uncertain.

Figure 9.15 Reaction checklist.

References

Black, T.H., 1983. The preparation and reactions of diazomethane, Aldrichim. Acta. 16, 3–10. http://www.sigmaaldrich.com/ifb/acta/v16/acta-vol16-1983.html.

Coyne, G.S., 2005. The Laboratory Companion, revised ed. Wiley-Interscience, New York, p. 35.

De Boer, T.J., Backer, H.J., 1956. Diazomethane. Org. Synth. 36, 16–19. http://dx.doi.org/10.15227/orgsyn.036.0016.

Dreisbach, R.R., Spencer, R.S., 1949. Infinite points of Cox chart families and dt/dP values at any pressure. Mathematical formulas. Ind. Eng. Chem. 41, 176–181. http://dx.doi.org/10.1021/ie50469a040.

Shanley, E.S., Greenspan, F.P., 1947. Highly concentrated hydrogen peroxide. Ind. Eng. Chem. 39, 1536–1543. http://dx.doi.org/10.1021/ie50456a010.

Following the Reaction

10

Chapter Outline

It is essential to follow the progress of the reaction whenever this is possible. Each time point in a reaction at which the chemist has some indication of what is happening is like a separate experimental result. The only other way to get this type of information would be to run replicate reactions for those periods of time, work them up, and determine the composition of the reaction mixture through analytical/spectroscopic techniques. Obviously, following one reaction is far easier than conducting and analyzing many. It is a good practice to save a small sample of starting materials as a reference and to obtain analytical/spectroscopic information on all reactants before the reaction is begun. Reactions that are new to the experimenter should be monitored at 5 min, 15 min, 30 min, 1 h, 2 h, 4 h, 8 h, and so forth. It is easy in a busy research laboratory to forget to take a data point to track a reaction. Chemists could take a lesson from the biology laboratory and use small, inexpensive electronic timers that can be programmed for multiple time points. Several analytical methods are available to follow reactions. The value of simple observations should also not be overlooked. Color or phase changes (precipitation or dissolution) may signal that a reaction has occurred. Temperature increases may occur because of exothermic reactions, a reason to include a thermometer that extends into the solution inside most reaction setups.

Handbook of Synthetic Organic Chemistry. http://dx.doi.org/10.1016/B978-0-12-809504-1.00010-8

10.1   Thin Layer Chromatography

Commercial thin layer chromatography (TLC) plates are generally large (20×20 cm) and come with several choices of backings, including glass and aluminum sheet. Many chemists prefer glass. While some glass TLC plates are prescored for breaking into particular sizes, this prevents the chemist from preparing plates customized to a particular use or from economizing by making smaller plates. It is better to obtain unscored plates and cut them as desired. Aluminum sheet TLC plates can be cut to size with a strong scissors or office paper cutter.

TLC plates come with a wide variety of adsorbents. The most common are silica gel 60 plates, which are optionally available with a fluorescent indicator that facilitates ultraviolet (UV) detection. Also available are alumina and reverse phase (see later for discussion of this concept) adsorbents.

10.1.1   Cutting Glass Thin Layer Chromatography Plates

Step 1—Use a guide rack or straightedge to ensure that parallel lines and appropriate spacing are maintained.
Step 2—Two types of glass cutters are available: the wheel type and the diamond type. The wheel type must be lubricated with a drop of mineral oil.
Step 3—Clean the glass surface of the plate with a tissue.
Step 4—Using the guide, score the glass with the cutter at desired intervals.
Step 5—Make all scores by moving the cutter in the same direction, pressing hard with a wheel cutter, firm with a diamond cutter.
Step 6—Turn the plate 90 degree and repeat.
Step 7—If a wheel cutter was used, clean the oil off plate with a tissue moistened with pentane.
Step 8—Break the scored plate into individual TLC plates by placing thumbs on either side of each scratch, protecting the adsorbent side of the plate with a paper towel, and bending up.

A video of this cutting technique and other techniques used in TLC is available here:

https://www.youtube.com/watch?v=ml58GCq078o.

10.1.2   Spotting Thin Layer Chromatography Plates

A pencil and straightedge are used to mark a line on the sorbent higher than the solvent level in the TLC chamber. Samples are then spotted along that line. TLC spotters are easily prepared by drawing out capillary tubes with two open ends over a flame. A little practice and experience should guide how long and narrow to draw the capillary. Almost all samples for TLC should be spotted from solutions of a milligram or so in a relatively volatile solvent. Spotting neat liquids will overload the plate. Nuclear magnetic resonance (NMR) samples are quite good for TLC spotting. Reaction mixtures are readily spotted simply with whatever is drawn into a capillary spotter. If the reaction is being conducted in a nonvolatile solvent (HMPA, DMF, DMSO), a large baseline spot will remain because these very polar solvents are poorly eluted. The

Figure 10.1 A filtering bell jar can be used to apply vacuum to a spotted thin layer chromatography plate to remove poorly volatile reaction solvents.
Photograph provided by Kimble/Kontes.

mobility of the reactants and products might also be affected because of dissolution of these solvents in the eluent as it moves past the baseline spot. However, it may still be possible to follow reactions in these solvents by TLC, especially if the solvent is at least mostly removed following spotting. This can be done by heating with a heat gun and/or pumping off inside a filtering bell jar (Fig. 10.1) with a rubber stopper. The TLC plate is held upright in a small beaker placed inside the bell.

It is good practice to save a small sample of each synthetic intermediate in vials for future TLC comparisons. Authentic samples of the nonvolatile reactants or other known compounds in the reaction should be spotted on the plate with the reaction mixture to provide more information concerning the identity of a particular spot. This spotting can be done in parallel tracks or even on the same track (called cospotting). The cospot aids the differentiation of reactants and products when they have similar R_Fs, and helps diagnose changes in the appearance of the TLC of a reactant when spotted from the reaction mixture. It occasionally happens that a starting material appears different on TLC when spotted from the reaction mixture rather than from a pure sample. This can mislead the chemist into thinking that the starting material has been consumed when it has not.

10.1.3   Thin Layer Chromatography Solvent Selection

TLC typically utilizes mixtures of a more polar and a less polar solvent so that eluent polarity can be readily adjusted by changes in the volume proportion of the two solvents. Normal-phase TLC involves a hydrophilic stationary phase (silica gel, alumina)

Table 10.1 Commonly Used Chromatography Solvent Mixtures

Hexanes/ethyl ether	Standard separations; nonpolar compounds
Hexanes/ethyl acetate	Standard separations; more polar compounds
Hexanes/tetrahydrofuran	Standard separations; yet more polar compounds
Hexanes/acetone	Standard separations; yet more polar compounds
Dichloromethane/methanol	Very polar compounds
Chloroform/MeOH/1% aqueous NH_3	For amines

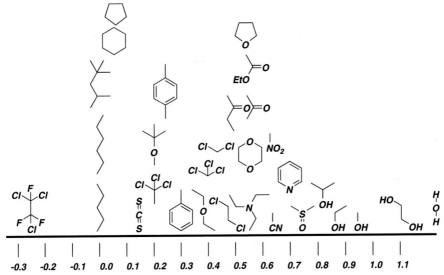

Figure 10.2 A graphical representation of the eluotropic series, which is eluting power in normal-phase chromatography.

such that more polar solvents have greater eluting power. Reverse-phase TLC involves a hydrophobic stationary phase (silica covalently bonded to hydrocarbon chains) such that less-polar solvents have greater eluting power.

Some solvent pairs commonly used for normal-phase chromatography in synthetic chemistry are summarized in Table 10.1. A series expressing the relative eluting power of a range of solvents in normal-phase chromatography is summarized graphically (Fig. 10.2). The domain of this plot of eluting power in this eluotropic series is solvent strength, expressed as ε.

When experimenting with different combinations of polar and nonpolar solvents to achieve the best separation, it can be difficult to estimate what the proportion should be in a new combination based on earlier results. A nomograph is provided (Fig. 10.3) that relates different solvent combinations and proportions.

Making substitutions in the traditional solvent systems can enhance their environmental friendliness, e.g., the chlorinated solvents (Chapter 7, Section 7.1) in the last two entries of Table 10.1. This consideration is even greater when analytical

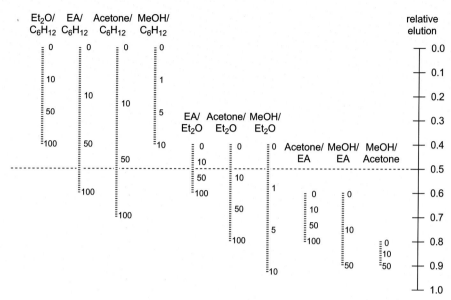

Figure 10.3 Nomograph relating the eluting power ε of different solvent combinations to their proportions. EA is ethyl acetate. Numbers adjacent to the scales refer to the volume % of the first-mentioned solvent. For example, switching from 20% acetone in cyclohexane to methanol in cyclohexane, one expects a comparable ε to be obtained with 5% methanol.

chromatography such as TLC or high performance liquid chromatography (HPLC) is extended to preparative purification (Chapter 14, Sections 14.2–3) because of the quantities of solvent used. The common dichloromethane/methanol system can be replaced by a ternary solvent system of isopropyl acetate, methanol, and heptane (Chardon et al., 2014). It is notable that heptane is increasingly being used as a substitute for hexanes in chromatography, as it is considered safer and less toxic, and only slightly less volatile. For ease in adjusting the eluting power of the ternary system, a 3:1 mixture of isopropyl acetate and methanol is used as the polar component and mixed in appropriate proportions with heptane. When separating acidic or basic compounds, acidic or basic modifiers, respectively, are added into the ternary mixture (2% by volume). In preparative separations, the proportion of the isopropyl acetate/methanol mixture in the eluting solvent should not exceed 80% to prevent the silica gel from being dissolved. Comparison of this new solvent system to the traditional methanol in dichloromethane shows that c. 70% 3:1 isopropyl acetate/methanol in heptane gives eluting power equivalent to 10% methanol in dichloromethane.

A long-studied topic in chromatography is the development of metrics to aid in the selection of solvent combinations for optimal separations. The most prominent of these was developed by Snyder (Snyder, 1978; Rutan et al., 1989) and is called the solvent selectivity triangle. Solvents were analyzed based on their physicochemical properties to derive three parameters, hydrogen bond basicity (X_e), hydrogen bond donation (X_d), and dipolarity (X_n). A graphic version of the Snyder solvent selectivity triangle is provided in Fig. 10.4.

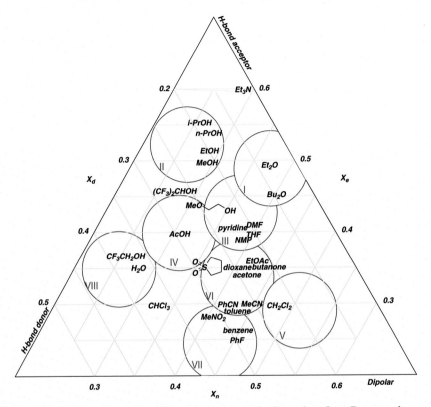

Figure 10.4 Snyder solvent selectivity triangle, prepared using values from Rutan et al. (1989) and Barwick (1997). Solvents now regarded as undesirable are still included in this figure to enable replacements for them to be suggested by proximity in the triangle.

The circles with red Roman numerals are solvent selectivity groups, which Snyder classified as follows:

I	Aliphatic ethers, trialkylamines
II	Aliphatic alcohols
III	Pyridines, amides, tetrahydrofuran
IV	Acetic acid, glycols, formamide
V	Dichloromethane
VIa	Aliphatic ketones, esters, dioxane
VIb	Sulfones, nitriles
VII	Aromatic hydrocarbons, nitro compounds
VIII	Fluoroalkanols, water

Note that in Snyder's formalism, these solvents would all serve as the polar component of a binary eluent, mixed with a nonpolar solvent (pentane, hexane, heptane, perfluorohexane (FC-72), cyclohexane, or carbon tetrachloride) to achieve a solvent of the eluting power required for the compound at hand. Also be aware that solvents

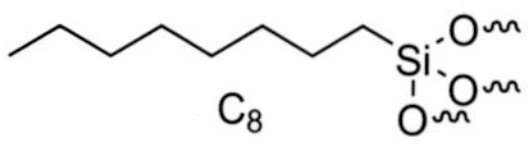

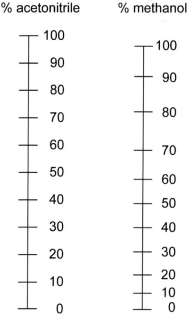

Figure 10.5 Reverse-phase silica gel has long-chain hydrocarbons chemically bonded to the silica backing (through a tris-siloxane group and the wavy bonds shown).

Figure 10.6 A nomograph relating the fraction of methanol or acetonitrile in water that provides equivalent mobility on a reverse-phase sorbent.

were included in this study without respect to their volatility, which is not important for analytical separations but is crucial for preparative chromatography (Chapter 14, Section 14.3).

Reverse-phase TLC involves a hydrophobic stationary phase (silica covalently bonded to C4, C8 (Fig. 10.5), or C16 hydrocarbon chains) such that less-polar solvents have greater eluting power. Mobile phases commonly used in reverse-phase TLC (or HPLC) include water/methanol and water/acetonitrile mixtures. The relationship between the relative eluting power of the two mixtures is not linear, but a nomograph (Fig. 10.6) can be used to relate the mobility of a compound in two different solvent systems.

For standard analytical TLC, the polarity of the eluting solvent should be adjusted such that the R_F of the compound of interest is around 0.5. This permits products that are either more or less polar to be easily observed. If the starting material spot is at the extreme bottom or top of the plate, products cannot be observed unless their polarity changes in the right direction. Of course, this rule may be violated if the polarity of the products relative to the starting material is already known.

Figure 10.7 Polar compounds can give unsymmetrical spots, called streaks or tails, on thin layer chromatography.

A problem common in the TLC of quite polar compounds is "streaking" (Fig. 10.7). It can be difficult to distinguish one compound that streaks from two compounds that have close R_Fs. A common remedy for streaking is inclusion in the eluting solvent of low levels of a component with the same chemical character as the compound being separated. That is, streaking of amines is reduced by ammonia or triethylamine in the eluent, and streaking of carboxylic acids is reduced by formic or acetic acid in the eluent.

10.1.4 Eluting Thin Layer Chromatography Plates

A wide variety of TLC chambers are available, and their design is not crucial, provided that they have a flat bottom and are narrow enough that the TLC plate cannot fall flat. Their primary function is simply to provide a thin layer of eluent solution to be drawn up the plate and an atmosphere saturated with eluent vapor. For the latter purpose, a filter paper is usually adhered to the wall of the chamber as a wick. After the solvent front has moved almost to the top of the adsorbent, the plate is removed and the solvent front is marked with a pencil for the determination of the R_Fs.

One problem that sometimes arises in TLC (and other adsorption chromatography) is that the compound may be unstable or otherwise chemically modified during the elution. A simple way to test for this possibility is two-dimensional analysis. A sample is spotted in one corner of a square TLC plate. It is eluted in one direction,

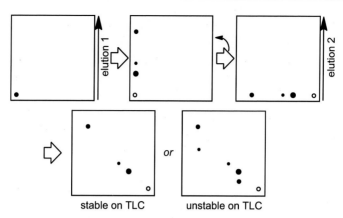

Figure 10.8 A two-dimensional thin layer chromatography experiment permits the identification of compounds that are unstable on silica gel.

dried, turned 90 degrees, and eluted again. It is then visualized. If the compounds have survived the TLC without modification, all spots should be found along the diagonal (Fig. 10.8). Off-diagonal spots indicate that modification has occurred during the elution. One can also perform two-dimensional TLC with a different solvent for the two elutions, potentially achieving greater separation than either of the solvents alone.

If compounds prove difficult to separate on TLC, multiple elutions may help. A solvent with about one-third of the polarity of that in which $R_F = 0.5$ is chosen. The plate is eluted and then dried before the next elution; this process may be repeated up to four times.

10.1.5 Visualizing Thin Layer Chromatography Plates

The first analysis of an eluted and dried TLC plate should be with a long-wavelength UV light ("black" light; Fig. 10.9). This method is nondestructive, whereas analyses with dips, sprays, or stains are irreversible. Compounds with UV chromophores that are fluorescent (primarily aromatic rings) may show up as vivid fluorescent spots under UV light, but this is not very common. If the TLC plates include a fluorescent indicator, compounds with UV chromophores that are not fluorescent (i.e., conjugated enones, dienes, or aromatic rings) can quench the fluorescence and show up as dark spots on a green background. Manufacturers typically use descriptors such as "silica gel F_{254}" to indicate TLC plates including a fluorescent substance with a 254 nm absorption.

A chamber containing iodine crystals will stain many organic compounds, and because of iodine's volatility, staining may be reversed simply by removing the plate from the chamber. Some indication of functional groups present may be obtained from iodine stains. Alkenes stain strongly, which is readily understood based on their reaction with halogens, and other unsaturated compounds are usually reliably stained. Saturated hydrocarbons, esters, ethers, and nitriles usually do not stain well, and many halogen compounds give a "negative" spot: a white zone on the slightly stained silica.

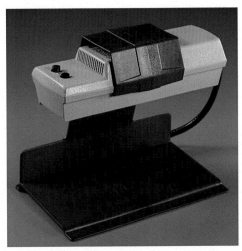

Figure 10.9 A long-wavelength UV lamp used for initial visualization of thin layer chromatography plates.
Photo provided by UVP LLC.

Other visualization methods use chemical sprays or dips. A wide range of stains are collected in Table 10.2. Recipes for the preparation of the various TLC stains are provided in Appendix 6. A few broadly used stains, such as ninhydrin, are commercially available. Dipping the plate in the hood into a beaker (covered with a watch glass) or a screw-cap jar filled with staining solution avoids toxic and corrosive chemical vapors from sprays, which were once popular. Using a forceps, the TLC plate is dipped into the solution up to the solvent front and removed. The back of the plate is cleaned with a tissue, and the plate is heated on a hotplate in the hood or with a heat gun. Spots usually develop in less than 2 min.

Caution must be exercised in interpreting the results of any TLC visualization, especially regarding the proportions of compounds in a mixture. Compounds that stain darkly or as large spots with a particular stain may be minor components and compounds that stain lightly or not at all may be major components. Indeed the whole purpose of some stains is to visualize some compounds selectively, most often based on the reaction of specific functional groups with reagents in the stain. The size of a spot on a TLC plate may have no direct relationship to the proportion of that component in the sample. The chemist may be seriously misled by attempting to "integrate" a TLC plate.

When applying any visualization method to a TLC plate, it is a good idea to circle the observed spots with a pencil so the exact position and size of the spots is preserved for following visualization steps. This information can aid in identifying compounds that are visualized by two different methods and help distinguish spots of compounds with close R_Fs but different staining properties. Circling the spots also deals with the common observation that spots fade over time. Finally, with the increased integration of technology into the laboratory, many workers are simply taking a digital photo of their TLC plates to be integrated into their electronic laboratory notebook or to print and include in their paper notebook. A now-archaic practice was to tape the actual TLC

Table 10.2 Stains for the Visualization of Spots on Thin Layer Chromatography Plates

Dip or Stain	Comments
Phosphomolybdic acid	General organic compound stain, plates must be heated to develop spots, does not distinguish different functional groups, spots are usually dark green on a light green background, alcohols typically give strong spots
Vanillin	General stain, amine, hydroxyl, and carbonyl compounds show strongly, spot colors are compound dependent, smells like cookies
Dinitrophenylhydrazine	General stain, identifies aldehydes and ketones by forming the corresponding hydrazones, which are usually yellow to orange
Cerium molybdate (Hanessian's stain)	General stain, usually requires vigorous heating, spots are often dark blue (though colors may vary) on a light blue or light green background
p-Anisaldehyde stain A	General stain, uses mild heating, sensitive to most functional groups, especially those that are nucleophilic, can be insensitive to alkenes, alkynes, and aromatic compounds, spot colors are variable on a light pink background
p-Anisaldehyde stain B	More specialized stain than A, uses mild heating, developed for terpenes
$KMnO_4$	Stain for functional groups sensitive to oxidation, alkenes/alkynes stain rapidly at room temperature as bright yellow spots on a purple background, other oxidizable functional groups (alcohols, amines, sulfides, and mercaptans) stain with gentle heating, their spots are usually yellow or light brown on a light purple or pink background, spots should be circled following visualization as the background becomes light brown with time
Ninhydrin	Selective for primary, secondary, and tertiary amines, amino sugars, classical detection for amino acids, take care not to get it on skin
Bromocresol green	Selective for functional groups such as carboxylic acids with pKa < 5, bright yellow spots on either a dark or light blue background, heating is not typically required
Ferric chloride	Selective for phenols
Ceric ammonium molybdate	Strong stain for hydroxy compounds
Ceric ammonium sulfate	Specifically developed for vinca alkaloids
Cerium sulfate	General stain, particularly effective for alkaloids
Ehrlich's reagent	Selective for indoles, amines, and alkaloids
Dragendorff–Munier stain	Selective for amines, and will stain amines of low reactivity that other stains do not

plate into the notebook, but this rarely gave enduring documentation; stains change and fade with time. Of course, the same can apply to printed digital photos; the most enduring form of a stained TLC plate is likely the electronic image. If an image is not part of the archiving used by the chemist, a simple line in the notebook with drawings for each spot and annotations of appearance and compound identity is still a workhorse method.

Depending on the particular stain used, compounds bearing different functional groups may or may not be distinguished based upon spot colors. Nevertheless, it may be worthwhile to note the colors of spots of particular compounds so that when similar chemistry is performed on related compounds, correlations of spot colors with functional groups may be discerned.

When following a reaction by TLC, it is generally assumed that any anionic or metalated intermediates in the reaction mixture will be protonated upon spotting onto the TLC plate (Eq. 10.1) and will migrate as the conjugate acids. In some cases, where reaction components interfere with TLC analysis, following a reaction may involve removing small aliquots and subjecting them to mini workups to obtain a solution that can be analyzed.

$$(10.1)$$

Evaluating the outcome of a TLC obviously entails measuring and recording the R_F of each spot, even if its molecular identity is known only provisionally. It is simple to compare R_Fs, but a more relevant measure of how their separation can be applied preparatively in column chromatography comes from the α factor or separation factor. The α factor is determined with respect to any pair of spots where spot 1 has the higher R_F (Eq. 10.2). The α factor is always >1 (a value of 1 corresponds to no separation), and better separations have larger α factors.

$$\alpha = \frac{R_F{}^1 (1 - R_F{}^2)}{R_F{}^2 (1 - R_F{}^1)} \qquad (10.2)$$

10.2 Gas Chromatography

The capillary column gas chromatograph with flame ionization detector (FID) is an excellent analytical tool, capable of analyzing products with sensitivity far greater than usually needed to follow organic reactions (i.e., ng). As the name implies, the column is a fused silica capillary 30–100 m in length whose inside surface is coated with the stationary phase. The primary basis of separation is the boiling point. Small samples of about 1 μL of solution are injected with a special GC syringe. Demonstrating the sensitivity of the instrument, much of the injected sample is split off to waste to prevent overloading the column. Crude reaction mixtures might be injected, but if salts or other nonvolatile compounds that can foul the injector are present, they are usually removed by filtration through a polar sorbent such as silica or alumina. Typically the column is temperature programmed, starting at a fairly low temperature (c. 100–150°C) and ramped up to high temperature to enable the analysis of a wide range of compounds. The mobile phase is a dry carrier gas, He. The FID uses H_2 and air to generate a flame that ionizes the column effluent, and it is this ionization that is detected. This detector thus responds to the ability of a compound to burn. One expects compounds that are isomers to have quite similar properties in this regard, and therefore their peak sizes should be proportional to their representation in the sample. Peaks for two compounds with different abilities to burn will not be directly reflected in their peak magnitudes (specifically, integrated area). If one wants to accurately measure relative amounts of two dissimilar compounds (e.g., for a kinetic study), it is necessary to determine the relative response of the detector to each compound. This is not often done for routine organic reaction monitoring. Also see the following method for internal standards (ISs) and calibration curves in HPLC.

10.3   High Performance Liquid Chromatography

Analytical HPLC is a more broadly applicable method to follow reactions than GC because some compounds are simply not thermally stable or volatile enough for GC. The volumes of sample analyzed are up to 40 µL per injection. A key element of the HPLC system (Fig. 10.10) is the injection valve, which includes a sample loop of typically 10 or 20 µL. In load mode, the loop is not in the fluid circuit entering the column, and the sample can be introduced via syringe, the liquid already in the loop being sent to waste. Upon switching the valve, the loop is incorporated into the fluid circuit and the sample can flow onto the column. This switching is recognized as the injection event by the data system recording the chromatogram. Often, injection leads a few minutes later to a wiggle or glitch in response, caused by the arrival of the solvent front at the detector.

Types of detection for HPLC include UV (both single wavelength and diode array detectors, which collect a full UV spectrum on the eluent at any point in time) and refractive index. Certainly, compounds with a recognized UV chromophore are

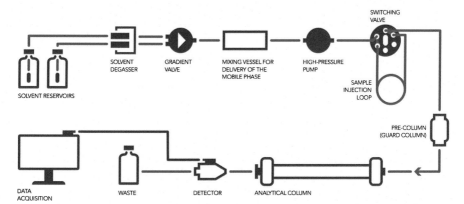

Figure 10.10 Schematic of a high performance liquid chromatography system.

readily detected, but at short enough wavelengths, even functional groups not considered very UV-active (such as amide carbonyl groups) are readily seen with a UV detector. Refractive index is a molecular property that saw significant use in organic chemistry when spectroscopic techniques were unavailable, but is hardly used today. Compounds with minimal functionality tend to have lower refractive index, while more functionalized and unsaturated compounds tend to have higher refractive index. Typical chromatographic solvents used with refractive index detectors are diethyl ether and hexanes, which have low refractive index. The strength of the detector signal for a specific compound will depend on the difference between its UV absorption or refractive index and that of the solvent. For this and other reasons, peak size is not directly related to the molar amount of substance in a peak. Separate measurements of response factors (the relationship of the amount of compound to the chromatographic signal) to calibrate the analysis are necessary if this information is desired (see later).

HPLC-MS instruments have become available at a price that makes them practical for individual departments or even laboratories to acquire. The reputation of mass spectrometry as a universal detector makes HPLC a powerful analytical tool. HPLC-MS consequently has become much more used to identify reaction products *during* an experimental run, rather than afterward. Separation of reaction mixtures by HPLC-MS tend to be run in reverse phase (see later) because the solvents used are more compatible with the MS part, especially electrospray ionization.

10.3.1   Normal Phase

Two major types of HPLC columns are used. Normal-phase HPLC is based on a polar sorbent, typically silica gel. This method is like conventional adsorption chromatography, with the advantage that the resolving power is much greater because the sorbent has much higher surface area and finer particle size, requiring high pressure to push the eluent through the column. The eluent can be a single solvent composition (isocratic) or involve an increase in the more polar component as the run proceeds (gradient). Like temperature programming in GC, elution gradients permit compounds of a wide variety of properties to be analyzed in a single HPLC run and ensures that all compounds loaded onto the column are eluted. Computer-controlled pumps mix the solvents in the correct proportion for the gradient. Many HPLCs also vary column temperature to optimize separations.

10.3.2   Reverse Phase

Reverse-phase HPLC is based on a nonpolar sorbent generated by derivatizing silica gel with hydrocarbon groups (Fig. 10.5). The eluting solvent is typically a mixture of water and either methanol or acetonitrile, and the gradient involves an increase in the organic component.

10.3.3   Quantitation of Reaction Products

When using HPLC to follow a reaction, product ratios and even an estimate of chemical yields can be determined, as contrasted with TLC. All that is required is a relationship between the size of a peak on the chromatogram and the amount of a compound. This information is available using ISs and calibration curves. They require only chromatograms of reference standards of known compound concentrations. Even though the latter are called curves, ideally there is a linear relationship between the concentration of compound and signal, as measured by the integrated area under a chromatographic peak. This value is reported by instrument software that analyzes the chromatogram. If UV detection is used, it is reasonable to assume linearity in detector response over several orders of magnitude in signal/concentration, in accord with the Beer–Lambert Law. This may be less true with other detection methods.

Just a few data points are needed to develop an HPLC calibration curve. A known volume (easily controlled based on the sample loop size) of a known concentration (determined based on volumetric preparation or from an analytical technique such as UV–Vis absorption) of compound is injected and the response is determined in area units. This gives a concentration/area relationship that should be consistent in replicate chromatographic runs. Linear regression of the response at different concentrations can provide a slope that is again concentration/area. Many software packages operating HPLC instruments facilitate the development of such calibrations.

ISs can be desirable to use when many manipulations are required before the sample can be analyzed. The presence of the IS allows material losses during these operations to be accounted for. For this method, a constant amount of a compound that is easily analyzed by the method used to follow the reaction is added to each reaction. The IS should be added in a concentration range 0.3–0.5 of the expected maximum product concentration, but this value may be adjusted if the response of the detector to the IS is significantly different from its response to the product. The IS *must* be totally inert to the reaction conditions studied. Molecules such as *tert*-butyl benzene and halogenated benzenes have seen a good deal of use as ISs. Some desirable characteristics of the IS include a significantly different retention time than is expected for any other peaks in the sample, a linear detector response, and ready commercial availability.

In an IS analysis of a reaction mixture, the ratio of the integrated area for the product peak and the IS peak is compared with the IS concentration to obtain the product concentration. For example, a standard mixture is prepared with a known concentration of the product (product standard, PS) and a known concentration of the IS. After HPLC analysis of this standard mixture, Eq. (10.3) is solved for F, a dimensionless constant. For experimental runs, the F so determined is entered into Eq. (10.4), along with the IS and product peak areas to determine the product concentration. HPLC software can also facilitate the development of IS calibrations.

$$F = \frac{\text{IS peak area} \times [\text{PS}]}{\text{PS peak area} \times [\text{IS}]} \qquad (10.3)$$

$$[\textbf{\textit{analyte}}] = F \times \frac{\text{product peak area} \times [\text{IS}]}{\text{IS peak area}} \qquad (10.4)$$

10.4   Nuclear Magnetic Resonance Spectroscopy

It is often possible to conduct a reaction on a relatively small scale in a solvent commonly used for NMR, such as d_6-benzene, d-chloroform, d_6-acetone, d_6-DMSO, d_3-acetonitrile, d_3-methanol, or D_2O, and observe reaction progress spectroscopically. Spectroscopic techniques that suppress solvent protons may also be used with nondeuterated solvents, or in some favorable cases it is not even necessary to use solvent suppression—the pertinent protons of reactants or products may still be readily observed (Hoye et al., 2004). For proton NMR, integrations of the peaks in the spectrum should be accurate to within about 10% of the actual compound ratios in the sample. Reactions that involve paramagnetic reagents or impurities or include suspended solids may be difficult to analyze by NMR because they give poor spectra. Particularly for the chemistry of air-sensitive transition metal complexes, where none of the chromatographic methods for following reactions is applicable, NMR is a prime means of reaction monitoring. Special NMR tubes that can be sealed with a flame under an inert atmosphere are often used to follow transition metal transformations.

References

Barwick, V.J., 1997. Strategies for solvent selection—a literature review. Trends Anal. Chem. 16, 293–309. http://dx.doi.org/10.1016/S0165-9936(97)00039-3.
Chardon, F.M., Blaquiere, N., Castanedo, G.M., Koenig, S.G., 2014. Development of a tripartite solvent blend for sustainable chromatography. Green Chem. 16, 4102–4105. http://dx.doi.org/10.1039/C4GC00884G.
Hoye, T.R., Eklov, B.M., Ryba, T.D., Voloshin, M., Yao, L.J., 2004. No-D NMR (no-deuterium proton NMR) spectroscopy: a simple yet powerful method for analyzing reaction and reagent solutions. Org. Lett. 6, 953–956. http://dx.doi.org/10.1021/ol049979+.
Rutan, S.C., Carr, P.W., Cheong, W.J., Park, J.H., Snyder, L.R., 1989. Re-evaluation of the solvent triangle and comparison to solvatochromic based scales of solvent strength and selectivity. J. Chromatogr. A 463, 21–37. http://dx.doi.org/10.1016/S0021-9673(01)84451-4.
Snyder, L.R., 1978. Classification of the solvent properties of common liquids. J. Chromatogr. Sci. 16, 223–234. http://dx.doi.org/10.1093/chromsci/16.6.223.

Working Up Reactions

11

Chapter Outline

The first consideration in working up a reaction is to remove any stopcock grease that was used on the joints of the reaction glassware. A tissue moistened with a solvent such as hexanes or dichloromethane does this well.

Having completed a reaction that included a magnetic stirring bar, it can be a challenge to pour the reaction mixture into a separatory funnel without also having the stir bar end up in the funnel. Special retrievers (magnets on inert plastic cords or rods) can be used to remove the stir bar from the reaction before transferring it to the funnel. Some dexterous chemists merely hold a powerful horseshoe magnet near the neck of the flask as they pour out the reaction mixture, catching the stir bar as it comes by.

A significant issue in planning the workup is whether solvent partitioning should be used. The main purpose of aqueous/organic partitioning is the removal of inorganic salts produced in a reaction, along with any water-soluble solvents, reagents, or by-products. If none of these is present, there is little point in solvent partitioning, and it may provide an opportunity for loss of water-soluble products. Evaporation of the

Handbook of Synthetic Organic Chemistry. http://dx.doi.org/10.1016/B978-0-12-809504-1.00011-X

solvent and direct purification has much to recommend it. In some cases compounds may not be able to survive aqueous workups, or one may simply choose not to do one. If aqueous/organic partitioning is needed and the reaction was conducted in a polar organic solvent, such as small alcohols, acetonitrile, or tetrahydrofuran, a worst-case scenario is that phase separation does not occur during partitioning because the polar organic is miscible with both water and the extraction solvent. To anticipate such problems, refer to the solvent miscibility chart provided in Appendix 3. Even if two phases are formed, often the presence of the polar organic solvent enables some product to dissolve in the aqueous phase and thereby be lost. In such cases it may be advisable to evaporate the polar solvent before partitioning the residue between water and a volatile organic solvent nicely immiscible with it.

For reactions using highly polar, poorly volatile solvents such as ethylene glycol, DMSO, DMF, and HMPA, getting rid of the solvent can be quite a challenge. It may be possible to remove a portion of them by evaporation before solvent partitioning, though high vacuum (and even warming) may be required, and not all products can tolerate this treatment. One good strategy is to partition the reaction mixture between an aqueous solution and a hydrocarbon solvent with which they are not miscible (Appendix 3), such that the polar reaction solvent prefers the aqueous phase. However, this ploy only works when the reaction product is not very polar and partitions predominantly into the hydrocarbon. DMSO is immiscible with diethyl ether, however, so ether can be used in extractive workups with DMSO. With a polar product that requires a more typical extraction solvent such as ethyl acetate, most of the DMF or DMSO can still be removed from the organic phase by five washes with twice the volume of water as the volume of DMF or DMSO. Some workers report using 5% aqueous LiCl in place of water to enhance removal of DMF. The strategy of using an aqueous solution and a hydrocarbon can also be applied to reactions in acetic acid, an approach far superior to using large quantities of hydroxide (or bicarbonate, with voluminous CO_2 evolution!) to extract the acetic acid. Of course, a base wash should still be used to ensure that all of the acetic acid has been removed from the organic phase.

Washing polar/acidic reagents away as an initial step in the workup is a profitable strategy. The condensation of lactic acid with acetone to give a dioxolanone is promoted by stoichiometric boron trifluoride etherate (Eq. 11.1). If this reaction were simply quenched with water, large quantities of fluoboric and hydrofluoric acids would be generated that would require large quantities of base to neutralize. Instead, the reaction mixture is diluted with ether and washed with aqueous sodium acetate. Some of the acid simply partitions to the aqueous phase, and some reacts with the acetate to form acetic acid, which partitions to the aqueous phase. The quantity and strength of the acid that must be washed out of the organic phase with base is thereby much reduced.

(11.1)

11.1 Solvent Extraction

The goal is to ensure that all of the desired product finds its way to the organic phase; as many and as much as possible of the by-products should go to the aqueous phase. Minimizing the volume of the aqueous phase, consistent with dissolving all of the salts, will maximize the organic product going to the organic phase. Extraction with solvents heavier than water makes the process operationally easier than the converse, but presents the problem that most such solvents are toxic chlorocarbons. Dichloromethane is the least noxious of the bunch. Other common extraction solvents are diethyl ether, hydrocarbons, and ethyl acetate. With molecules particularly difficult to extract from the aqueous phase, mixed solvents may prove helpful. For example, a mixture of 10–15% isopropanol in a chlorocarbon is one of the most polar organic solutions that is immiscible with water. It is quite effective in extracting polar organics such as phosphonates or heterocycles (Fig. 11.1). To ensure easy phase separation when using such solvent mixtures, it is essential to consider densities such that one phase is significantly more or less dense than the other. Solvent miscibility and therefore utility in phase partitioning (including chromatographic methods) is summarized in Appendix 3, including a chart of the miscibility of common organic solvents referred to earlier.

The aqueous phase typically is extracted a few times with equal volumes of the organic solvent and the combined organic phases are washed. The wash solutions vary with the reaction: $NaHCO_3$ solution and 10% NaOH solution will remove acidic compounds, depending on their pK_A, while 5% HCl is used to remove bases. $NaHSO_3$ and $Na_2S_2O_3$ are common reductants used to destroy excess oxidants. Saturated $CuSO_4$ solution is very specific for removing pyridine. It forms a blue-purple complex and acts as its own indicator, since the $CuSO_4$ solution is deeply colored when pyridine is present. This method also works with amines used commonly as bases, such as triethylamine. If the reaction involved a polar organic solvent such as DMF or DMSO that may have been carried along to this final stage, washing with five times the volume of the DMF or DMSO that was used in the reaction should ensure that none persists. Final washes can be done with brine to remove any traces of water from the organic phase.

The lessons of solvent extraction from beginning organic chemistry classes should not be forgotten by the aspiring synthetic organic chemist. While modern separation tools are amazing and powerful, one should not be seduced by the sophistication of the new. All too often, novices use a high-cost, labor-intensive technique such as

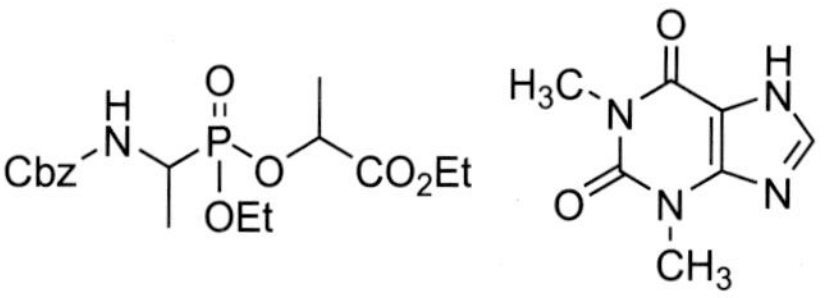

Figure 11.1 Examples of quite polar compounds that are extracted from aqueous reaction mixtures using $CHCl_3$/isopropanol mixtures.

chromatography to separate two components in a reaction mixture (e.g., a phenol and a neutral) that could be easily separated by solvent extraction. A nice example of how compounds can be manipulated in solvent extraction via their acidity is seen in the hydrolysis of a 2,6-dimethylphenyl ester (Eq. 11.2). Alkali initially produces the conjugate bases of both the acid and the phenol in an aqueous phase. If acidified with strong acid, both bases would be protonated and would be extracted into the organic phase. However, if CO_2 (in the form of ground dry ice) is added to the alkaline solution, it effectively converts the hydroxide to carbonate, which is only basic enough to ionize the acid, not the phenol. The phenol is removed by extraction into an organic phase, and the aqueous phase is acidified with HCl to recover the acid by extraction into a different organic phase.

$$(11.2)$$

Problems can and often do arise in solvent extraction. For example, when using extraction with base to move an acid to the aqueous phase, its salt may be quite insoluble even though it is ionic. This event is difficult to anticipate; likely the only way to avoid this problem is to do a small-scale test extraction before the whole reaction mixture is extracted. Strategies to overcome the problem lie mainly in trying different cations (sodium→potassium, lithium, or ammonium). If this fails, the alternatives are to isolate the crystalline salt, or simply acidify the whole mixture to go back to ground zero. Keep in mind that salts of organic compounds are essentially soaps, and can facilitate the extraction of neutral organics into the aqueous phase. It may be possible to back-extract the aqueous phase to pull these neutrals back into an organic solvent before the aqueous extracts are combined for neutralization.

Likely the foremost difficulty in extractions is the formation of emulsions. Emulsions are most likely to occur with solvents of high viscosity and a low surface tension, and when there is a small density difference between the phases. These property data for a selection of solvents are collected in Table 11.1. Some are volatile solvents that would be used for extraction, and some are reaction solvents that may be present in the mixture. If it is the reaction solvent, not the extraction solvent, that seems to be creating the difficulty with the emulsion, an obvious approach is to change the problem solvent or remove it, by evaporation if possible, before extraction. Other tactics to deal with an emulsion include centrifugation, or filtration of the whole *gemische* through a pad of Celite (trade name for diatomaceous earth). Celite is a filter aid composed primarily of silica, which can remove very fine particles that would otherwise clog a filter. Ultrasonication (in an Erlenmeyer flask) might also be worth a try. Increasing

Table 11.1 **Selected Solvent Properties (at 25°C) That Can Affect Emulsion Formation**

Solvent	Density (g/mL)	Surface Tension (mN/m[a])	Viscosity (mPa·s[b])
Acetonitrile	0.786	28.66	0.369
Chloroform	1.492	26.67	0.537
Dichloromethane	1.325	27.20	0.413
Diethyl ether	0.706	16.65	0.224
Dimethylsulfoxide	1.100	42.92	1.987
Ethanol	0.789	21.97	1.074
Ethyl acetate	0.902	23.39	0.423
Hexane	0.672	17.89	0.300
Methanol	0.791	22.07	0.544
Pentane	0.626	15.49	0.224
Pyridine	0.978	36.56	0.879
Tetrahydrofuran	0.889	26.80	0.456
Toluene	0.865	27.93	0.560
Water	1.000	71.99	0.890

[a]mN/m was formerly dyne/cm.
[b]mPa·s was formerly cP.

the difference in density between the phases or increasing their surface tension are other approaches. This can be done by dissolving a neutral electrolyte such as Na_2SO_4 or NaCl in the aqueous phase or by adding a more polar, less-dense solvent (ether, ethyl acetate, pentane) to the organic phase.

11.2 Drying Organic Solutions

After an aqueous extraction, the organic phase will include some dissolved water, so its drying is necessary before solvent evaporation. Four agents are commonly used for this purpose: $MgSO_4$, Na_2SO_4, K_2CO_3, and molecular sieves. A complete listing of desiccants is provided in Table 11.2. The choice can be based on the following considerations. It is easy to tell when $MgSO_4$ has fully dried an organic solution because dry $MgSO_4$ is fluffy but the hydrate is clumpy. A powerful drying agent such as $MgSO_4$ may be necessary for the somewhat polar and hydrophilic ether and ethyl acetate. While $MgSO_4$ is fast, it is also acidic, and acidic enough to remove ethylene ketals. For example, the apparently selective formation of the ketal of the saturated ketone in androstenedione (Eq. 11.3) is in fact ketalization of both carbonyl groups and a rapid hydrolysis of the α,β-unsaturated ketal promoted by the $MgSO_4$ drying agent. If a compound is acid sensitive, K_2CO_3 can be substituted for $MgSO_4$. K_2CO_3 is granular and free flowing when dry but clumpy when wet, and of course is slightly basic. If a compound bears base-sensitive groups, such as acetate esters, and a protic solvent is present, removal of the acetate might be expected. Na_2SO_4 likewise is granular and free flowing when dry but clumpy

Table 11.2 Chemical Desiccants

Agent	Used With	Not Used With	wt% Water[a]	Mechanism
Al$_2$O$_3$	Hydrocarbons		20	Adsorption
CaCl$_2$	Halides, hydrocarbons, esters, ethers	Alcohols, aldehydes, amides, amines, ketones	30	Hydration
CaO	Alcohols, amines, NH$_3$(g)	Acidic compounds, esters	30	Chemisorption
MgO	Alcohols, aldehydes, amines, hydrocarbons	Acidic compounds	50	Hydration
MgSO$_4$ (anhydrous)	Most compounds, including acids, aldehydes, esters, ketones, nitriles	None	c. 50	Hydration
P$_2$O$_5$	Anhydrides, ethers, halides, hydrocarbons, nitriles	Acids, alcohols, amines, ketones	50	Chemisorption
K$_2$CO$_3$ (anhydrous)	Alcohols, amines, esters, ketones, nitriles	Acids, phenols	20	Hydration
KOH (pellet)	Amines	Acids, phenols, esters	NA	Hydration
Silica gel	Most compounds		20	Adsorption
Na$_2$SO$_4$ (anhydrous, granular)	Acids, aldehydes, halides, ketones		120	Hydration

[a]g/100 g desiccant.

when wet. It is also quite neutral, but it tends to be slower than either MgSO$_4$ or K$_2$CO$_3$, such that overnight drying of organic solutions is often required unless they are quite nonpolar and nonhydrophilic such as dichloromethane. Molecular sieves can also be used, but they offer none of the self-indicating features of any of the chemical drying agents. Following drying, filtration into a flask suitable for vacuum (round-bottom or pear-shaped) prepares for evaporation of the organic solvent.

$$(11.3)$$

11.3   Specialized Workups

11.3.1   Reactions Producing Triphenylphosphine Oxide

Such reactions include Wittig and Mitsunobu reactions. The following method will work only if the product is relatively nonpolar. The reaction mixture is concentrated, the residue is suspended in pentane or hexanes and a minimum amount of ether, and this solution is loaded onto a short "plug" (5 cm in height) of silica gel in a chromatography column. The product is eluted with ether, leaving most of the phosphine oxide on the silica gel. Sometimes it is necessary to repeat this procedure two to three times.

11.3.2   Reactions Involving Boron Compounds

Many boron compounds and boron-containing residues (e.g., from Suzuki coupling, NaBH$_4$ or B$_2$H$_6$ reduction, or hydroboration) can be removed from a reaction mixture by evaporating it repeatedly from MeOH. This process forms (MeO)$_3$B, which is volatile and therefore also removed.

11.3.3   Reactions Involving Copper Salts

Such reactions include those using organocuprates. Quench the reaction with a saturated solution of aqueous NH$_4$Cl and allow the mixture to stir for a few hours while open to the atmosphere. This will oxidize all cuprous salts to the cupric oxidation state, which is much more reactive to ligand exchange. This will permit complexation of the cupric ion by ammonia, as indicated by the dark blue color of the solution.

11.3.4   Reactions Involving Aluminum Reagents

Such reactions classically involve LiAlH$_4$, but the method described here may be more broadly applied. The aluminum hydroxide formed upon simple aqueous quenching can create impressive emulsions. This material can be dissolved in aqueous acid, but not all reaction products can tolerate this, and product losses into the aqueous phase can be substantial. In particular, when an amine is prepared by LiAlH$_4$ reduction, it will be converted to its salt if an acidic workup is used and partition completely to the aqueous phase. A method that avoids adding large amounts of acid to work up LiAlH$_4$ reductions is called the "Fieser" or N,N,3N workup (Micovic and Mihailiovic, 1953). The reaction mixture is carefully quenched (H$_2$ evolution!) with 1 mL of H$_2$O for each gram of LiAlH$_4$ used. When performed on a significant scale, the mixture should be diluted with the reaction solvent and cooled in an ice bath for this step. After a few minutes of stirring, 1 mL of 15% NaOH for each gram of LiAlH$_4$ is added, and after another few minutes of stirring, 3 mL of H$_2$O for each gram of LiAlH$_4$ is added. This sequence converts the aluminate salts into alumina, which should be an easily filtered, granular material. No further aqueous workup is necessary. A drying agent can be added if desired, and filtration and evaporation provide the product.

Applying this approach to other aluminum reagents (alkoxy aluminum hydrides, aluminum alkyls) is simply based on the molar amount of aluminum in the reagent rather than the mass. The reaction mixture is quenched (H$_2$ evolution!) with 0.04 mL of H$_2$O for each mmol of aluminum reagent. After a few minutes of stirring, 0.04 mL of 15% NaOH for each mmol of aluminum reagent is added, and after another few minutes of stirring, 0.12 mL of H$_2$O for each mmol of aluminum reagent is added.

If these approaches fail, it is possible to remove aluminate salts by a traditional method, complexation with a 10% solution of Rochelle's salt (potassium sodium tartrate tetrahydrate).

11.3.5 Reactions Involving Tin Reagents

Such reactions include the free radical reduction of halides. Sodium fluoride solution is useful for the workup of reactions of Bu$_3$SnH, which works by the production of insoluble fluoride salts that are removed by filtration (Caddick et al., 1993). Stille couplings also produce trialkyl tin compounds and may be purified by similar methods.

11.4 Destroying Reagents

Many residual reagents must be destroyed once the reaction is complete. This can be because they are reactive or because they are toxic and require specific hazardous waste disposal procedures. The process can be obvious and intrinsic to the workup procedure, like quenching a Grignard reagent. It could also be a treatment required for phases produced in a solvent partitioning scheme, and is especially important when using any particularly hazardous materials. In adjusting the quantities in the following procedures to the reaction at hand, provision should be made for a completely failed reaction, requiring *all* of the added reagent to be destroyed, even when it was intended to be consumed during the experiment. Information beyond that provided here can be obtained from a comprehensive text on the topic (Lunn and Sansone, 2012).

11.4.1 Metal Azides

Solutions of NaN_3 should NEVER be put into aqueous waste disposal streams. It can be destroyed by treatment with nitrous acid, which produces nitrous oxide gas. Because of the anesthetic properties of nitrous oxide, it is important to perform this procedure (like essentially *all* chemical processes) in a well-functioning fume hood. A 200 mL aqueous solution containing 10 g of NaN_3 is treated with 15 g sodium nitrite in 75 mL water. To this solution is slowly added $4 M H_2SO_4$ until it is acidic to pH paper. The resultant solution can be checked for the presence of residual azide as follows. A few drops of the solution are added to aqueous KI (10% w/v). A drop of 1M HCl is added to ensure an acidic pH, and a drop of starch solution is added as an indicator. If the characteristic purple color of starch/iodine develops, excess nitrous acid is present and the azide destruction should be complete. If color does not develop, nitrous acid is not in excess and additional sodium nitrite solution should be added. This procedure can also be performed on a buffered solution.

11.4.2 H_2O_2, Organic Peroxides, Peracids, and N-Chlorosuccinimide

It is important to destroy any residual peroxide reagents before evaporating solutions containing them. Their hazards increase greatly with concentration, and what is safe to handle in dilute solution can become dangerous when the solvent is removed. This procedure should be scaled to account for the actual amount and strength of oxidant in the experiment at hand. To destroy 5 mL of 30% hydrogen peroxide, 100 mL of a 10% aqueous solution of sodium metabisulfite is added. For organic oxidants, this volume of metabisulfite solution should be able to reduce c. 5 g of peroxide. Starch/iodine test paper should be used to verify by the absence of color development that all of the peroxide has been consumed.

11.4.3 Chromium Oxidant Salts (Aqueous)

This procedure must be performed carefully, as it can be exothermic on larger reaction scales. Sodium metabisulfite solution is added until the solution tests negative with starch/iodine. A solution of c. 5 g $Mg(OH)_2$ is added, the mixture is stirred for 1 h, and it is allowed to stand overnight. The liquid deposits large amounts of Cr^{+3} sediment, which must be treated in accordance with hazardous waste protocols. The supernatant is decanted through a paper filter and the filtrate is checked with starch/iodine to be certain it is not oxidizing. If it is oxidizing, it must be acidified with H_2SO_4 and the protocol repeated. The resulting filtrate should contain only traces of Cr^{+3}.

11.4.4 Hydrogen Fluoride

Many fluoride-containing reagents are used in synthetic chemistry, often in the removal of silicon-based protecting groups. Their fluoride ions are converted to the quite poisonous HF at appropriate pHs. A general method to eliminate fluoride ion is conversion with calcium salts (such as the oxide) to the nontoxic and essentially insoluble CaF_2. A commonly used fluoride-containing reagent is $BF_3 \cdot OEt_2$ (boron trifluoride etherate), which is converted to boric acid and HF upon aqueous hydrolysis. For each 1 mL of $BF_3 \cdot OEt_2$ (25 mmol of fluoride), add 2.5 g of CaO dissolved in 100 mL water. After stirring the solution for 18 h, allow the CaF_2 to settle and remove it by filtration.

11.4.5 Cyanide

Recognizing that cyanide ion is poisonous and is converted to the quite poisonous gas HCN at appropriate pHs, extreme caution must be exercised during the use of cyanide in reactions, in accordance with safety data sheets and standard operating procedures. Once the reaction is complete, residual cyanide is inactivated by oxidation to cyanate. The concentration of cyanide is adjusted to the equivalent of 25 mg/mL aqueous NaCN. For each volume of solution, one volume of NaOH and two volumes of laundry bleach (5.25% NaOCl) are added. This process is exothermic and must be performed incrementally. After stirring for 3 h, completion of the reaction is indicated by testing for the presence of oxidant using starch/iodine, and the solution is neutralized. A similar procedure can be applied to the destruction of cyanogen bromide (BrCN). If the cyanogen bromide is in an organic solvent, its concentration is adjusted to 20 mg/mL and aqueous NaOH and NaOCl added as above.

11.4.6 Sodium Cyanoborohydride

Because it contains a cyanide ion that is not consumed during its reactions, sodium cyanoborohydride must be treated just like cyanide. Each gram of reagent is dissolved in 10 mL water and 200 mL of 5.25% NaOCl (laundry bleach) is added with stirring.

11.4.7 Acid Halides/Anhydrides

These commonly used acylating agents are reactive and corrosive. They are converted to carboxylate salts by hydrolysis with 2.5 M NaOH, which may require water-miscible organic cosolvents such as tetrahydrofuran to dissolve them.

11.4.8 Alkylating Agents

A wide variety of alkylating reagents are used to add alkyl groups in synthetic chemistry, include alkyl halides, sulfonate esters, and epoxides. These compounds are characteristically toxic, some acutely, related to their ability to alkylate essential biological molecules. For simple alkyl halides such as methyl iodide and benzyl chloride or bromide, 1 mL of the halide is dissolved in 25 mL of 4.5 M ethanolic KOH. The solution is heated at reflux for 2 h with stirring, then cooled. The mixture is diluted with 100 mL water, the phases are separated, and the aqueous phase is carefully neutralized. α-Haloethers such as chloromethyl methyl ether (MOM-Cl) and chloromethyl benzyl ether (BOM-Cl) are more reactive than the simple alkyl halides. If they are in water-miscible organic solutions, they are diluted to 50 mg/mL and an equal volume of 6% NH_4OH is added. The solution is allowed to stand for 3 h and neutralized with H_2SO_4. If they are in water-immiscible organic solutions, an equal volume of 33% NH_4OH is added and the two-phase system is shaken on a mechanical shaker for 3 h. Glassware that has been used in reactions of haloethers can be decontaminated by rinsing with acetone and combining those acetone washes with solutions to be inactivated as earlier. Dimethyl sulfate (and related agents like methyl triflate) is highly reactive, which explains its acute toxicity but also its easier destruction. If it is in a water-miscible organic solution, shake with 1.5 M NH_4OH until homogeneous, c. 15 min. In water-immiscible organic solutions, rapidly stir with 1.5 M NH_4OH for 24 h.

11.4.9 Thiols and Sulfides

Destruction of these compounds is necessitated not by their hazards but by their stench. Commercial hypochlorite laundry bleach (5.25% NaOCl when fresh) is used to oxidize them. A fourfold molar excess should be sufficient to complete the oxidation to a nonvolatile sulfonic acid or sulfone. On appreciable scale, oxidation is indicated by an exotherm. Temperature should be controlled at c. 50°C. Generation of sulfonic acid can cause the pH to drop from the basic bleach solution. Sodium hydroxide should be added if the pH is below 6 to prevent loss of the oxidant under acidic conditions.

References

Caddick, S., Aboutayab, K., West, R., 1993. An intramolecular radical cyclisation approach to fused [1,2-a]indoles. Synlett 231–232. http://dx.doi.org/10.1055/s-1993-22414.

Lunn, G., Sansone, E.B., 2012. Destruction of Hazardous Chemicals in the Laboratory, third ed. Wiley, Hoboken.

Micovic, V.M., Mihailiovic, M.L., 1953. The reduction of acid amides with lithium aluminum hydride. J. Org. Chem. 18, 1190–1200. http://dx.doi.org/10.1021/jo50015a017.

Evaporation

12

The rotary evaporator (Fig. 12.1) is an essential piece of equipment in most organic laboratories. Commonly, vacuum is provided by a water aspirator (or by other mild vacuum sources), and the flask is rotated by an electric motor. This provides a thin film of solution for evaporation, hopefully preventing the solution from bumping. The flask rotates in a temperature-controlled water bath, which provides the heat of vaporization. Little heat may be needed to evaporate highly volatile solvents such as ether, whereas heat is definitely required for solvents such as toluene. If the reaction product is somewhat volatile, warming the bath should be avoided to prevent compound loss.

It is essential that a glassware bulb (Fig. 12.2) be inserted between the flask and the evaporator to make provision for bumping of the solution being evaporated. If a bump does occur, the solution is trapped in the bulb, enabling it to be transferred back into the flask. Without the bulb, the solution would be drawn up into the evaporator itself, where it could mix with condensed solvents, be contaminated by other compounds that have been earlier evaporated in the equipment, or be lost altogether. This bulb naturally must be rinsed with

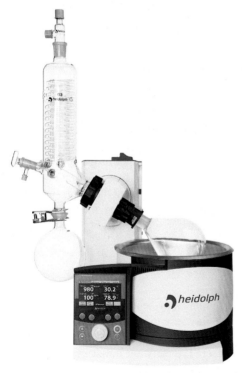

Figure 12.1 The rotary evaporator is essential in any synthetic organic chemistry lab and is the scientific instrument most fascinating to lay reporters. It is invariably the backdrop for any television or print news story on chemistry.

Handbook of Synthetic Organic Chemistry. http://dx.doi.org/10.1016/B978-0-12-809504-1.00012-1

Figure 12.2 The rotovap bulb prevents bumps from solvent in the evaporating flask from being carried into the rotary evaporator.

solvent after evaporation of each sample. The flasks used on a rotary evaporator may be of many sizes and types, but typically round-bottomed or pear-shaped flasks are used. Flat-bottomed flasks are not recommended because they are less robust to vacuum.

The evaporator generally has a condenser or trap (circulating cold water or dry ice) that can condense evaporated solvents to prevent their being drawn into the waste water or vacuum source and/or to potentially permit solvent recovery and recycling.

Few beginning chemists have avoided the following embarrassing scenario: A flask was fitted onto the joint on the rotary evaporator, the vacuum and motor were turned on, and the chemist walked away, having forgotten to close the vent to the vacuum system and wait for the vacuum to take hold to secure the flask. The friction fit of the flask on the joint of the glassware trap may have held it for a time, but eventually it slipped off into the water bath. As frustrating as this situation is, all is not lost. It is absolutely possible to perform solvent extraction on the water in the rotovap bath to recover at least some of the product. The yield obtained in today's reaction is not something one could quote for publication, however. Considering the effort that goes into carrying synthetic intermediates through several chemical steps, the chemist simply cannot afford to allow this loss of material without some effort at recovery. This experience also serves as a useful lesson that most never forget.

A video of rotovap techniques is available here:

https://www.youtube.com/watch?v=3DQj4dibr78.

The size of the flask used in evaporation should ideally be at least twice the volume of the solvent. Often, this means a fairly large flask even if the amount of product is 100 mg or less. Such large flasks may not even fit onto the analytical balance used to obtain an accurate mass of the reaction product. Typically, an initial evaporation of the large volume is followed by transfer of the residue into a smaller, tared flask. This transfer is conveniently done using a long Pasteur pipette and a rubber bulb. A few microliters of a nicely volatile solvent such as ether or dichloromethane are used to dissolve the residue by squirting the solvent down the walls of the flask while taking care to avoid the ground-glass joint. The resulting solution is then drawn into the pipette and transferred into the smaller flask. Repetition of this process once or twice more should ensure quantitative transfer.

The chemist is sometimes faced with the task of transferring a viscous material into a small container, such as a vial. Dissolving it in a solvent is necessary, but evaporation in the vial is a challenge. An adapter can be fitted with a rubber stopper that snugly fits into the mouth of the vial, and rotary evaporation can be applied.

In some instances, it will be impossible to remove the solvent using a rotary evaporator because the reaction product is too volatile. This applies to low molecular weight compounds as well as fluorinated compounds whose vapor pressure is unusually high. The only real alternative to the rotovap in such cases is fractional distillation using a Vigreux column. Recall from Chapter 3 that such a column can separate compounds differing in boiling point by 30°C, so a solvent with an appropriate boiling point must be chosen. After all of the solvent that can be removed in this way has distilled out, the still head temperature will drop and the apparatus is allowed to cool to ambient temperature. Residual solvent in the apparatus will return to the still pot, leaving a mixture of the solvent, product, and any higher boiling components of the crude reaction mixture.

Centrifugal evaporators (Fig. 12.3) are relatively new to the chemistry laboratory, having made their way over from the biology laboratory. They were initially designed for the vacuum evaporation of ethanolic DNA solutions held in plastic tubes. Many different rotors are now available to accept many types of laboratory containers in large numbers. Centrifugal force keeps the liquid within each container and prevents bumping losses. This instrument is essential to high-throughput, parallel approaches to organic synthesis.

Rotary evaporation is rarely sufficient to remove all traces of solvent, especially if a higher boiling solvent such as toluene or a solvent with a high heat of vaporization such as *tert*-butanol is present. It is always advisable to place the evaporation flask under high vacuum (<1 Torr). With solvents that are more difficult to evaporate or viscous samples, this high vacuum may still prove insufficient to remove the last of it, even when kept under vacuum overnight. A quick application of heat from a heat gun (Fig. 12.4) will frequently show the telltale signs of evaporation in the form of bubbling. At this stage the product should show little residual solvent by NMR.

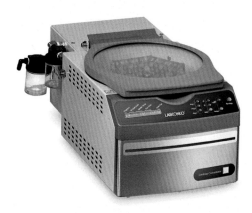

Figure 12.3 A centrifugal evaporator (colloquially, a SpeedVac). Tomy Seiko and subsidiaries.
© Labconco Corporation. Used with permission.

Figure 12.4 A heat gun.

It is a good practice to save a TLC-sized sample of the crude reaction mixture (after evaporation but before any purification) and to record its NMR spectrum. These data can assist in understanding a purification by column chromatography, since TLC and NMR data will be obtained on each fraction. It sometimes appears by TLC that the major component of the reaction mixture has been isolated, but the NMR of the crude reaction mixture reveals that it is in fact a minor component and the major component is still to be found. The ratios of starting material to product (percent conversion) are available from crude NMR data, as are the ratios of product(s) to by-product(s) and the ratios of any isomers present. When there is a possibility of isomer formation and their ratio is important, it should be determined before any fractionation of the reaction mixture. It is possible that one of the isomers could be selectively lost during purification, thereby perturbing the actual ratio that was obtained in the reaction. Crude TLC and NMR data are also important to understanding all the products generated during a reaction, not just the major product, which is helpful during reaction optimization.

It is also important to obtain the mass of the crude reaction product after evaporation. While a yield cannot be calculated until the product has been purified, knowing how much of the mass that was added to the reaction mixture has been returned in the crude product ("the mass balance") can be very useful in troubleshooting reactions. If the mass balance is low, for example, the workup may have been poorly designed to recover the desired product. If the mass balance is far above the expected value, the solvent may have polymerized. The mass balance is simply determined by evaporating the crude reaction mixture in a tared flask. Since evaporation flasks are either round-bottomed or pear-shaped, they must be prevented from tipping over or rolling across the bench. Cork rings or rubber filter adapters are available in sizes appropriate to all size of flasks and do an excellent job. A visit to any synthesis lab will reveal benches with many flasks containing products. It is crucial to label these flasks to avoid mix-ups. Adhesive labels could be used, but they change the tare of the flask. Better are tags with short loops of string that can be wrapped around the neck of the flask.

Vacuum Systems

13

Chapter Outline

13.1   Vacuum Sources

Vacuums are obtained in the laboratory in at least three ways: "house" vacuum, water aspirators, and vacuum pumps. The strength of the house vacuum is defined locally. Water aspirators should be capable of providing a vacuum of about the vapor pressure of water. This is around 24 Torr at room temperature; colder water will translate to a better vacuum. This pressure is quite adequate for many operations, such as rotary evaporation and some distillations. A trap bottle should be inserted between the aspirator and the apparatus, consisting of a thick-walled Erlenmeyer filter flask bearing a rubber stopper with a hose connection and a stopcock (for releasing the vacuum). Its role is to prevent water from entering the apparatus in case of a backup at the aspirator. Long (15 cm) tubing extending downward from the aspirator is necessary to obtain optimum vacuum. Some chemists object to aspirators because they waste water. For the green chemist, aspirator pumps that recirculate the water are available.

A common type of laboratory vacuum pump uses an electric motor and belt-driven rotary oil pump (e.g., Welch 1400 Duo-Seal; Fig. 13.1). These pumps are capable of vacuums of 0.1 Torr (100 μm) or better when operated with manifolds and other apparatus. They are used for vacuum lines, inert atmosphere boxes, distillations, and so on. Another type of vacuum pump has direct drive between the motor and the oil pump and is fully integrated. These are commonly used with instruments such as mass spectrometers because of their low maintenance. For the vacuum required in solvent delivery systems (Section 7.2), diaphragm pumps are available that are low maintenance and resistant to organic solvents.

13.2   Vacuum Manifolds

Working at reduced pressure is significantly easier when a glass vacuum manifold is used. It may be used with (or integrated with, as shown in Fig. 13.2) the inert

Handbook of Synthetic Organic Chemistry. http://dx.doi.org/10.1016/B978-0-12-809504-1.00013-3

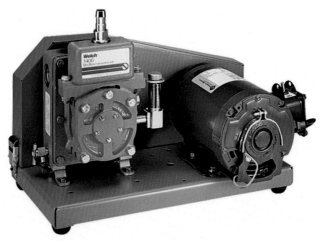

Figure 13.1 A conventional laboratory oil vacuum pump.
Courtesy of Welch, by Gardner Denver Inc.

Figure 13.2 A glass double manifold for simultaneous manipulation of the atmosphere in a
vessel, including vacuum and delivery of an inert gas.

gas manifold discussed earlier. Components of the system often include a pump
connection, a cold trap (to condense organic vapors, preventing them from being
drawn into the pump), a central chamber with several branching stopcocks (best
are hollow vacuum stopcocks whose interior can be evacuated to keep the stopcock seated; Fig. 13.3), and a pressure gauge. It is essential in any vacuum system
that one stopcock always remains available to the ambient atmosphere to permit
release of the vacuum. A vacuum pump must be vented to the atmosphere when
turning it off; otherwise, oil will be drawn into the manifold system. The cryogens
dry ice/isopropanol or liquid nitrogen are used to cool the trap.

Safety Note

If liquid N_2 is used to cool the trap, it should not be applied until the system has mostly been evacuated. Otherwise, liquid oxygen (O_2) can be condensed in the trap, which can be explosive if organic compounds are also present. Liquid oxygen is indicated by its blue color. The ideal trap design has stopcocks both before and after the trap as well as a stopcock to vent it to the atmosphere. This arrangement allows the trap to be brought to atmospheric pressure for emptying without venting the rest of the vacuum system. It is likewise hazardous to expose a liquid nitrogen-cooled trap to an argon atmosphere; substantial quantities of liquid argon can be condensed, and when the liquid nitrogen is removed, the liquefied argon will boil, leading to a rapid increase in volume and pressure. If the vacuum system is closed to the atmosphere at this time, its explosion is certain, and explosion is even possible when the system is open to the atmosphere.

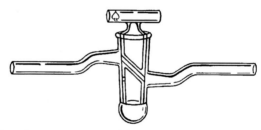

Figure 13.3 A hollow plug vacuum stopcock is the stopcock most resistant to leakage.

When a vacuum system is not performing well, the following items should be checked. These are also good maintenance practices. Although there are no general rules for how often these measures need to be taken, monthly checks of these items are advised.

1. Change the pump oil. Run the pump for 15 min to warm the oil before draining it. Add new oil slowly to give it a chance to flow until it appears at the correct level, typically in a sight glass. If the removed oil seems particularly dirty, it may be worth using a "flushing oil" first. These materials are actually hydrocarbons that are less viscous than oil. Run the pump for 15 min against a closed system, drain the flushing oil, and replace it with the normal (high-vacuum) oil. Repeat the foregoing with high-vacuum oil. Simple replacement of the high-vacuum oil is the normal maintenance procedure.
2. Check that all ground-glass joints are fully seated and well greased with a good high-vacuum grease such as Apiezon M. The same goes for stopcocks, except the grease should be Apiezon N.
3. Minimize the lengths of hoses and the numbers of joints and stopcocks.
4. Fill the cold trap with a coolant such as dry ice/isopropanol or liquid N_2.

Safety Note

All glassware used in vacuum work or containers that are under vacuum must be securely and adequately taped or shielded with plastic mesh to restrain flying glass in case of an implosion or other accident. Never use glassware that is etched, cracked, star-cracked, chipped, nicked, or scratched, which makes it more prone to breakage under vacuum or pressure. Use glassware that has been repaired only upon the advice of the master glassblower who performed the work. All glassware under vacuum should be protected with plastic safety shields.

13.3   Vacuum Gauges

It should be emphasized that while it is crucial that the pressure be reported for some operations such as vacuum distillation, there are significant errors in most simple measurements of vacuum.

At least two types of McLeod gauges (swivel and tipping) are available to measure vacuum/pressure. They have a rest position (when they are exposed to the vacuum) and a reading position. With the swivel gauge (Fig. 13.4), the mercury reservoir is down at rest. The gauge has both linear and nonlinear scales, with the latter for higher accuracy at pressures less than 1 Torr. To read the pressure, the reservoir is swiveled forward and up past the horizontal, until the mercury level in the closed capillary reaches the lowest line as shown (Fig. 13.5). Pressure is read from the level of the mercury in the open capillary. If the pressure is less than 1 Torr, the nonlinear scale [graduated in microns

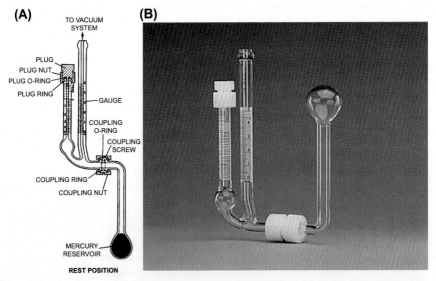

Figure 13.4 The swivel McLeod pressure gauge, in schematic (A) and photographic (B) forms. © Sigma–Aldrich Co. LLC.

(∞, 10^{-3} Torr)] is used as follows: continue to swivel the reservoir up until it reaches vertical, then carefully adjust the mercury level in the open capillary until it is at the top line. The level of the mercury column in the closed capillary provides the pressure.

Tipping McLeod gauges (Fig. 13.6) measure pressures from around 5 Torr to 5 μm. In the rest position the glass portion is turned 90 degrees to the right from the position

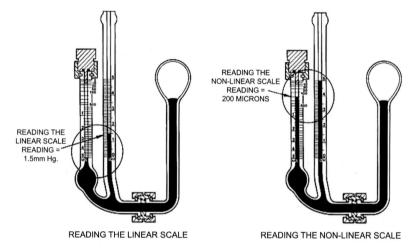

READING THE LINEAR SCALE   READING THE NON-LINEAR SCALE

Figure 13.5 Reading the swivel McLeod gauge on the linear and nonlinear scales. © Sigma–Aldrich Co. LLC.

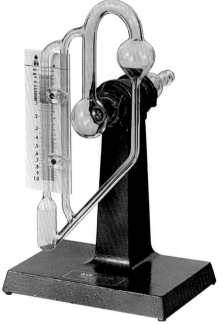

Figure 13.6 A tipping McLeod pressure gauge. © Sigma-Aldrich Co. LLC.

shown, with mercury filling the pear-shaped reservoir. To read the pressure, the glass portion is rotated to the left (to the position shown) and mercury enters the closed arm. The pressure is given on the scale on the closed arm.

Mercury manometers (Fig. 13.7) are simpler and cheaper than McLeod gauges, though they cannot measure pressures lower than a few Torr. They are often used in conjunction with aspirators. Like many manometers, pressure readings are based on the difference in the heights (in mm, translated to Torr) of the inner and outer mercury columns.

Safety Note

Manometers are the most likely source of larger mercury spills in the laboratory, but broken thermometers are the more frequent source. Mercury metal is a toxic, volatile liquid. Mercury vapor is readily absorbed by inhalation and can also pass through intact skin. Mercury is highly hazardous when inhaled or when it remains on the skin for more than a short period of time, commonly resulting in penetration to the central nervous system and mercury poisoning (long-term neurological and kidney damage). Elemental mercury is not well absorbed by the gastrointestinal tract.

Mercury spills generate an enormous number of tiny droplets that are easily dispersed, and special vacuum equipment must be used to clean up these spills. The safety office or other environmental services can be called to safely and properly clean up the spill. Make sure someone stays near (but not in) the spill zone to keep people away. Do not touch or attempt to wipe up spilled mercury. Do not spread sulfur powder on the spill. Take all measures necessary to prevent mercury from entering sink or floor drains.

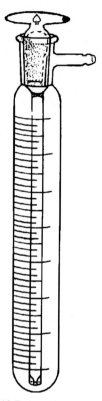

Figure 13.7 A mercury manometer.

There are also modern alternatives to classical mercury manometers—electronic pressure meters known as baratrons or ionization gauges, which have specific pressure ranges in which they can operate.

Purification of Products

14

Chapter Outline

If the methods described so far have not produced a single pure compound, as evidenced by the number of spots on TLC, the number of peaks on GC or HPLC, or an NMR spectrum, purification must be undertaken.

14.1 Distillation

Kugelrohr distillation serves only to separate nonvolatile materials from volatiles, and is always performed under reduced pressure, which keeps the glassware assembled while in use. This technique can be performed on a relatively small scale (hundreds of milligrams) without serious material losses. It is also sometimes called bulb-to-bulb distillation (Fig. 14.1). One "bulb" (round-bottom flask) is held in an electrically heated oven and is attached through a hole in the oven wall to an outside bulb (Fig. 14.2). This latter bulb is attached by a hose adapter and rubber tubing to a pneumatic drive motor that reciprocally rotates the glassware, like a windshield wiper, to prevent "bumping" (sudden violent eruptions of vapor that carry liquid over into the receiver). The motor drives a hollow shaft through which the vacuum is pulled. The external bulb is cooled (with an ice/H_2O or dry ice/acetone bath) to

Handbook of Synthetic Organic Chemistry. http://dx.doi.org/10.1016/B978-0-12-809504-1.00014-5

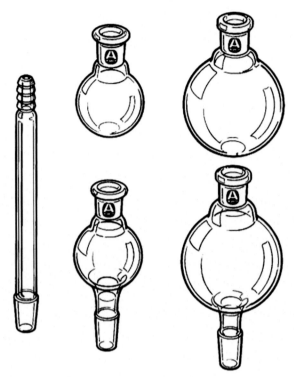

Figure 14.1 Glassware for bulb-to-bulb distillation.
© Sigma–Aldrich Co. LLC.

(A)   **(B)**

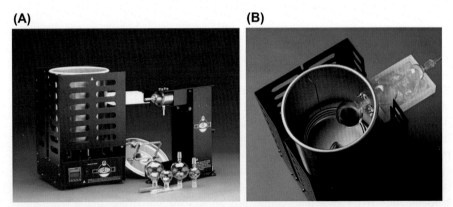

Figure 14.2 A Kugelrohr distillation oven. (A) Side view; (B) top view.
© Sigma–Aldrich Co. LLC.

condense the vapor. Once the compound has distilled, it may be removed from the bulb by pouring it out (if it is free flowing) or washing it out with ether.

The short-path distillation method described for the purification of reagents may also be applicable to product purification, provided that the scale is toward the larger, 25 mmol end of the spectrum that is the main focus of this book. The number of compounds for which distillation is applicable is somewhat limited, but when it is, distillation can be an excellent purification method. Compounds purified by chromatography may remain tenaciously colored despite appearing quite pure by other criteria, whereas distilled compounds are most often water-white (provided they have no chromophore).

14.2 Silica Gel Chromatography

Silica gel chromatography is by far the most widely used method of product purification because it is so general and because it is a natural outgrowth of the analysis of reaction mixtures by TLC. For the most part, solvent mixtures that are effective for TLC can also be used for column chromatography, with some adjustment of proportions, mostly to reduce the R_F. One exception is that acetone should not be used in the elution of a preparative column because it easily undergoes aldol condensation to give 4-methylpenten-2-one.

For many years, column chromatography was performed with gravity elution that might require hours. Automated fraction collection was essential to permit columns to run while the chemist performed other tasks. This all changed with the development of flash chromatography (Still et al., 1978), which has many virtues, chief among them being much faster separations. Gas pressure is used to push eluent through silica gel with a small pore size; separation is generally complete in minutes, meaning that fractions can be collected manually in test tubes in a rack. It is thus a "poor man's" HPLC.

14.3 Flash Column Chromatography

Since its original development, many proper and improper modifications have been made in or incorporated into the practice of flash chromatography by novice synthetic chemists. It is therefore worthwhile to return to the original publication to review the first principles of this widely used method. First, it uses 40–63 µm silica gel, which had been used for thin-layer chromatography. Still et al. (1978) recommend selecting an eluting solvent that gives an R_F on TLC of the target compound of 0.35. It is important to note the difference between this recommendation and that for analytical TLC. He also recommends column diameters for particular separation scales (Table 14.1) and that the pressure be adjusted so that the solvent head drops at the rate of 5 cm/min. Still claimed the ability to separate compounds whose R_Fs differ by 0.15, and in some favorable cases, 0.1. Past studies on gravity flow columns have reported separations of compounds whose R_Fs differ by 0.01. So, one limitation of the flash chromatography

Table 14.1 Recommended Parameters for Flash Column Chromatography at Different Scales[a]

Sample Loading (mg)	Column Diameter (mm)	Fraction Size (mL)
100	10	5
400	20	10
800	30	20
1600	40	30

[a]Still et al. (1978).

Table 14.2 Polar Solvent Components and Eluotropic Strength for Flash Chromatography Method Development

Solvent	Strength ε
Ethanol	0.65
Acetone	0.53
Acetonitrile	0.52
Ethyl acetate	0.36
tert-Butylmethyl ether	0.32
Dichloromethane	0.30

method is reduced resolving power. Having experienced the rapid separation of a flash column, however, chemists are unlikely to return to gravity flow columns to achieve separation of close-running compounds. Preparative chromatography instruments available today are much more likely to be used for this task.

While a solvent system for analytical TLC should already be known, it may still be necessary to develop a new system for optimal flash chromatography separation of the crude reaction mixture. Additional materials may be present that had not been considered in development of the TLC solvent. Methods have been recommended for selecting a flash chromatography solvent based on binary mixtures of less polar (heptane or toluene) and more polar components (Dubant and Mathews, 2009). The polar components and their relative eluting power are given in Table 14.2. Obviously, for solvents with larger ε, their proportion must be reduced to achieve an R_F comparable to binary systems with a polar component of lower eluting strength.

The mass of silica gel appropriate for a particular flash chromatography separation can be determined based on TLC data of the mixture, using a simple algorithm that also suggests the best fraction volume (Fair and Kormos, 2008). This algorithm is available in spreadsheet form and is downloadable from the companion website for this book. The

Figure 14.3 A flash chromatography column, reservoir, and air control valve.

recommended mass also gives an estimate for the amount of solvent required; for each g of silica gel, 10 times that volume in milliliter of eluting solvent should be needed.

Flash chromatography uses a column with a Teflon stopcock and a solvent reservoir (Fig. 14.3). The reservoir ideally stores sufficient eluent to complete the chromatographic separation so that the pressure need not be released to replenish the solvent. The column should be run from start to finish in one session; pauses permit molecules to diffuse, broadening bands and decreasing separation efficiency. A cotton ball in the exit stopcock is used to hold in the packing; this is preferable to a frit, which introduces dead volume and can clog. If the size of the cotton ball is chosen correctly, it is possible to use air pressure to blow it into and lodge it in the exit stopcock. If this does not work, a long glass rod must be used to push the cotton ball into the stopcock. A small amount of washed sand is added to form a thin, level layer onto which the silica gel will pack.

There are two ways in which the column can be loaded with silica gel. Traditionally, gravity columns are slurry packed—silica gel is mixed in a flask or beaker with sufficient eluting solvent to wet and suspend all of it. This mixing releases heat that can evaporate some of the more volatile solvents. After stirring to remove air bubbles, a slurry is formed that is added into the column via a funnel. The stopcock is opened so that solvent flows off the column as the silica gel settles. Repeated addition of eluting solvent to the flask or beaker is necessary to wash all of the silica gel from it onto the column. This method can also be used to pack a flash column, but has the disadvantages of being messy and time-consuming. The alternative is dry packing, described following.

A measured amount of silica gel appropriate for the separation is carefully added to the column via a powder funnel. The column can be tapped if needed to form a level top layer. A valve consisting of a Teflon screw valve, a 24/40 joint, and a gas inlet is placed on the top of the column and attached to a source of compressed air. Some chemists firmly attach the valve to the column and adjust the flow rate with the air pressure and the screw valve. Others close the screw valve completely, turn on the compressed air, and use the force with which the valve is pressed onto the top of the column (by hand) to control the pressure and therefore the flow rate. The reservoir is loaded with an initial volume of eluting solvent, and it is pushed through the column. As expected, the mixing of dry silica gel with the solvent produces heat that cracks the silica gel bed, but this is not a problem because continued pushing of solvent through the bed drives out bubbles and packs the silica gel uniformly. After the column is equilibrated and the solvent is pushed through until it is just at the top of the silica packing, the column is topped by a thin layer of washed sand and the sample is added. It can be added in a minimal volume of eluting solvent by pipetting it onto the top of the column, or it can be dissolved in a volatile solvent that is mixed with silica gel and evaporated to give a dry powder that is poured onto the top of the column.

A video of flash chromatography techniques is available here:

https://www.youtube.com/watch?v=fF1gXUvyGb4

Flash chromatography can be conducted on a very small scale, and is more useful than other more classical small-scale chromatographic techniques such as preparative TLC. A Pasteur pipette can be used as a flash column, for example. Glass wool is pressed into its narrow neck and covered with a layer of sand. Silica gel is added to about half the height of the main barrel, the pipette is clamped to a rack, and eluting solvent is added above. Tygon tubing connected to a compressed air source is slipped over the top of the pipette to provide pressure. Fractions are collected in a rack of small vials.

It is possible to perform flash chromatography with air-sensitive compounds (Kremer and Helquist, 1984). Special glassware, particularly a sophisticated flow controller, is used that permits the whole operation to be carried out with and in an inert atmosphere. A glassblower's skills will likely be needed to fabricate such equipment.

Commonly, silica gel used in a flash column separation is discarded after use in accord with local chemical waste procedures. While it is possible to recycle it, the cost in time and effort is comparable to the cost of new silica gel.

14.4   Gradients

Despite some similarities to HPLC, column chromatography differs from it in that continuous solvent gradients are rarely used, as they are difficult to generate. In some cases step gradient elution may be used, for example, involving several column volumes of 10% solvent A in solvent B, next 30% A in B, then 70% A in B. In flash chromatography, this technique requires release of the pressure to change solvents in the reservoir, whose inconvenience may inhibit the use of gradients.

14.5 Special Adsorbents

14.5.1 Triethylamine-Treated Silica Gel

Compounds that are sensitive to the acidic character of conventional silica gel can sometimes be separated if the silica gel has been pretreated with a deactivating agent. An eluent that includes 1% by volume of triethylamine may be used in the initial wetting and equilibration stage preparing for flash chromatography. Column elution may be conducted with or without triethylamine in the eluent. Laboratory mates will appreciate elution without triethylamine because of its base odor. Examples of sensitive compounds that can be purified using this strategy include nucleoside phosphoramidites (Fig. 14.4).

14.5.2 Oxalic Acid–Coated Silica Gel

This material was developed for the purification of particularly sensitive quinones (Fig. 14.5) (Yamamoto et al., 1976), and has been used in flash chromatography (Pirrung et al., 2002). It is prepared by suspension of silica gel in 0.1 N oxalic acid overnight, filtration, washing with H_2O, and drying in an oven at 100°C overnight.

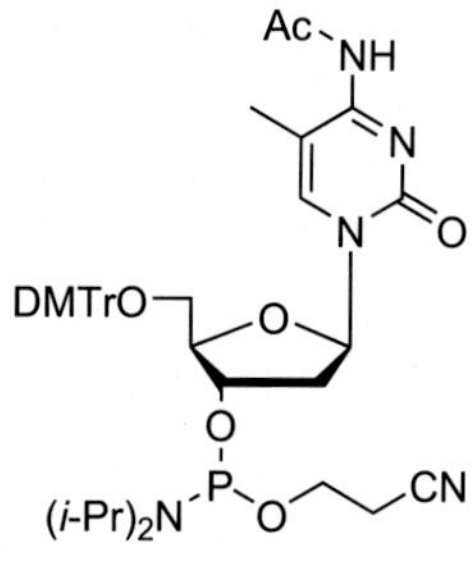

Figure 14.4 A nucleoside phosphoramidite can be purified by chromatography on triethylamine-treated silica gel.

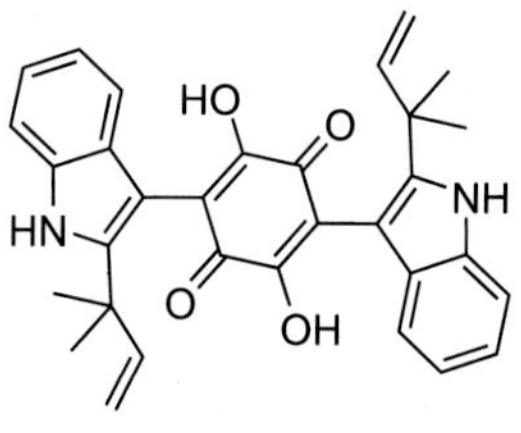

demethylasterriquinone B4

Figure 14.5 A quinone whose purification by chromatography requires oxalic acid–coated silica gel to avoid decomposition.

14.5.3   Silver Nitrate–Impregnated Silica Gel

This sorbent was developed long ago to address the issue that hydrocarbons generally have little affinity for polar silica gel (Williams and Mander, 2001; Mander and Williams, 2016). Therefore hydrocarbons migrate with a high R_F on conventional silica gel sorbents regardless of the eluting solvent, and no separation of hydrocarbons can be achieved. An approach was sought to increase the interaction of hydrocarbons with the stationary phase and form the basis of separations. Many hydrocarbons, including those that are naturally occurring, include C=C bonds, and it was known that such bonds have a high affinity for metal ions, particularly silver. The idea then was that silver ion would be added to the stationary phase to increase the retention of alkenes. Not only did this achieve their separation from alkanes, differently substituted alkenes could be separated from one another based on the strength of their retention by the stationary phase, as alkene affinity toward silver ion is dependent on substitution. It is thus possible to separate the turpentine constituents α-pinene and β-pinene (Fig. 14.6). Protocols for preparation of silica columns including silver ion were therefore developed. These primarily involve dissolving silver nitrate in methanol or acetonitrile, adding it to the silica gel, and evaporation and oven drying. These manipulations can be a bit challenging because of the photosensitivity of silver, however. They also require special care in handling by the chemist, since silver ion stains skin black.

Access to silver nitrate–impregnated TLC plates is essential to making the decision to use silver nitrate chromatography. They must also be used in selection of solvents and analysis of column fractions. Getting and using these plates is somewhat problematic because the photoreactivity of silver ion makes advance preparation and storage of silver-impregnated silica gel difficult. To avoid the need for preparation of homemade TLC plates that include silver, methods have been developed to modify commercial TLC plates with silver (Ratnayake, 2004). Soaking them in 12.5% (wt/vol) aqueous AgNO₃, or allowing this aqueous solution to be drawn up the plate as if one was performing TLC, adds the required metal ion. Spraying of the AgNO₃ solution onto the plate is less desirable because it is not uniform. Water deactivates the silica gel, so the silver TLC plates must then be reactivated by heating at 80–110°C for 1–2h. These operations and storage of the plates are performed with the exclusion of light to the extent possible.

14.5.4   Other Sorbents

Other stationary phases that can be used for column chromatography include florisil, which is magnesium silicate, and alumina. Broadly speaking, their adsorptive

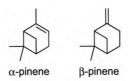

α-pinene   β-pinene

Figure 14.6 The terpenes α-pinene and β-pinene can be easily separated only on silver nitrate-impregnated silica gel.

properties are greater than silica gel. Reverse phase sorbents were mentioned for following reactions (Chapter 10, Sections 10.1, 10.3); they can also be used in flash chromatography (with rather polar and not very volatile eluting solvents, however, such as aqueous/organic mixtures). Since reverse phase sorbents must be prepared by chemical derivatization of regular silica gel, they are much more expensive than silica gel. As a consequence they are less practical with scale and typically cannot be discarded after one use. They must be washed with a strongly eluting solvent (a high proportion of the organic component) to remove all possible compounds remaining from the last separation.

14.6   Preparative Gas Chromatography

Packed column GCs, as contrasted with capillary columns, may still be found in some laboratories. With relatively large diameter columns (¼-inch steel tubing), these instruments can accept large amounts (up to $50 \mu L$) per injection. With a thermal conductivity detector, the analysis is nondestructive, allowing preparative separations. This basically involves monitoring the recorder to identify when a desired compound is eluting. A 3 mm glass tube or a specially made collector is attached at the exit port and the vapor flowing off the GC column is condensed. Some compounds condense better when the collector is cooled (ice or dry ice), while others form aerosols with cooling and are better collected in a warmer container (wrapped with aluminum foil). After collection, one end of the glass tube may be sealed and the liquid centrifuged to the sealed end. This sample is now ready for mass spectrometry or combustion analysis.

14.7   Precipitation

Literally, this just means a compound coming out of solution. A simple example of the use of this process is in purification of side-chain protected polypeptides obtained from solid-phase synthesis. The peptide is fairly soluble in solvents such as DMF, but precipitates upon the addition of water as an antisolvent for the peptide, and with which DMF is miscible. Isolation in this way is fairly attractive because it is usually impractical to remove DMF by evaporation. The precipitated solid can be isolated by filtration, decanting, or centrifugation. Mostly, material produced via precipitation has not formed a proper crystal lattice and is considered amorphous.

14.8   Trituration

This technique involves grinding a solid under a solvent. It is applied to solids that are agglomerates of desired compound(s) as well as impurities, and might be used if efforts to purify by crystallization have failed. The strategy is to select a solvent in which the desired compound is insoluble and the impurities are soluble. Grinding is performed simply to ensure that all surfaces of the solid have adequate contact with the solvent so the impurities can dissolve. It is performed with a spatula or glass rod and may be repeated following decanting of the solvent.

14.9   Crystallization

Crystallization is a process of growing crystals from solution, not simply the transition from solute to solid that is reflected in precipitation. It is recrystallization if the substance was already crystalline, but any solid or liquid can be crystallized. An attraction of crystallization as a purification method is that it can be scaled up easily, and indeed because of physical losses intrinsic to the method, it often works better on larger scales than small.

Lessons on crystallization are taught early in a chemistry career: choose a solvent in which the desired compound is insoluble at lower temperature but soluble at higher temperature (and, ideally, impurities are always soluble), warm to dissolve the crude material, then cool to allow crystallization. This seems simple enough, but the devil is in the details. Foremost among the details is what solvent to use. Much of crystallization uses common laboratory solvents, such as water, alcohols, acetone, ethyl acetate, cyclohexane, and toluene; it is also wise to recall the "like dissolves like" dictum.

Solvent selection can be made by analogy—if similar compounds have been crystallized from a particular solvent, that provides good teaching about where to begin with the current compound. A resource where the crystal and/or solubility properties of a large range of compounds is collected is the section Physical Constants of Organic Compounds in the *CRC Handbook of Chemistry and Physics*; it can be accessed in electronic form for easy searching. Older editions of the *CRC Handbook* offer two solubility listings for some compounds, one for hot solvent and one for cold. This information is obviously very pertinent to crystallization. One challenge in looking to older literature for crystallization solvents is that sometimes those used in the past are considered too hazardous to use under current chemical safety regimes, i.e., benzene, chloroform, carbon disulfide, or carbon tetrachloride. Possible replacements for these solvents were discussed in Chapter 7, Section 7.1.

If the foregoing solvent information is not available for the compound of interest, the only choice is empirical screening. Solvents and mixtures can be used. Reasonably volatile solvents are preferred, which facilitates freeing crystals from the crystallization solvent. They need not be as volatile as reaction/extraction solvents, however, where it is essential the solvent be readily evaporated to recover products. Solvents with boiling points up to c. 150°C can be used for crystallization. For crystallization trials, though, it is better to begin with solvents more volatile than this. If a compound dissolves in a solvent but does not show differential solubility with temperature, that solvent cannot be used for crystallization. It must then be evaporated to recover the precious sample of compound for trials of other solvents.

What solvents should be tried? A selection of solvents with a statistically designed range of properties has been recommended for growing crystals of different types (polymorphs or habits) for molecules already known to be crystalline

(Allesø et al., 2008). It includes acetonitrile, benzotrifluoride, decane, ethyl acetate, hexane, methanol, 2-methoxyethanol, nitromethane, perfluorohexane, toluene, and trifluoroethanol. Referring to Chapter 7, Section 7.1 for solvent properties may also be useful in selecting solvents to investigate. The solubility of a compound in candidate crystallization solvents is tested at an initial concentration of 100 mg/mL. If it is not fully soluble upon warming, solvent is added to increase the volume by 50%. If the compound is still not soluble, further additions of half-volumes of solvent can be made. If it has not dissolved when the initial volume has been tripled, it is considered sparingly soluble in that solvent and another must be chosen.

Mixed solvents are commonly used in crystallization, most often one that dissolves the solute readily and one in which it is only sparingly soluble. It is essential that the solvents be miscible, however. Refer to Appendix 3 for information on this property. There is also some analogy between solvent mixtures used for chromatography and those used for crystallization. Standard practice is to dissolve the solute in the better solvent at its boiling point and add the weaker solvent until cloudiness is detected, indicating that solute is coming out of solution. A small amount of the better solvent is added to clarify the solution, and it is allowed to cool for crystallization. Common solvent pairs are water/ethanol, ethyl acetate/hexanes, and ethyl ether/hexanes. In mixed solvent systems, it sometimes appears that the better solvent dominates the solute properties of the mixture at higher temperature and the poorer does so at lower temperature.

Low-melting solids, which can be considered anything with a melting point below 100°C, can pose a particular challenge in crystallization. First, there is a correlation between the melting point of a compound and the ease of crystallization. If one does not know that a new compound is a low-melting solid, crystallization may not even be attempted. When it is attempted, low-melting solids have the annoying tendency, upon cooling the crystallization solution, to separate as an oil that will not crystallize. Colloquially, this is "oiling out." It occurs because the solution reaches saturation at a temperature above the melting point of the solid. This problem can be addressed by using more solvent so that saturation is not reached until the temperature is below the melting point.

A frustrating aspect of crystallization is that sometimes a molecule will appear pure by standard measures, such as spectroscopy, but it still will not crystallize, even when it is known to be crystalline. Abandoning crystallization for purification by chromatography, the chemist then finds the chromatographed compound crystallizes readily. Presumably, chromatography removed small amounts of impurities that impeded crystallization. This scenario is more likely with low-melting solids.

Crystal growth requires nucleation, the formation of crystal seeds, which can be triggered by traditional methods such as scratching the glass with a rod, or by modern ones such as ultrasound (see Chapter 9, Section 9.7.4). It is sometimes possible to promote nucleation by freezing the crystallization solution solid with a cryogen. Nucleation is endothermic in most cases so it is not impeded by such cooling, and the lower solubility at low temperature may help. If seed crystals can be obtained in this way, it is best to warm the solution for crystallization to control the rate. This works well because crystal growth is exothermic in most cases.

It is crucial to closely observe crystallization attempts until nucleation begins. If seed crystals are not visible, something must be changed. When they do appear, crystal growth can be controlled by temperature and cooling rate, to be modified empirically for best results. The best crystals typically are formed by slow cooling without agitation, so an arrangement should be developed where temperature is controlled and (mechanical) vibration is absent. It is better to crystallize from a dilute solution in which the solute is less soluble than a concentrated solution in which it is more soluble. Solutions that are too concentrated are more likely to give glasses. When using a hydrophilic crystallization solvent (or even diethyl ether), one trick to aiding nucleation is to use a nonglass vessel. Hydrophilic solvents flow up glass walls as they evaporate and the seed crystal formation that occurs there is outside of the solution, where it needs to be to nucleate crystallization.

For the crystallization of compounds on the small scale commonly used at the forefront of multistep synthesis, use of a Craig tube is recommended to minimize physical losses. It enables the separation of crystals from crystallization solvents without filtration. Training in the use of these tubes is common in undergraduate teaching laboratories using microscale techniques.

14.10   Yields

It will not take a beginning synthetic chemist long to appreciate the importance of yield when performing multistep syntheses. The amount of starting material available for the next step is dependent on the yield of the last step, rather than the accessibility of a commercially available compound. This means that yield is one of the prime considerations for evaluation of a synthetic method or a particular transformation, thereby providing workers an incentive to "put their best foot forward" when reporting yields. Yield is also a source of endless frustration when one cannot attain the yield desired/ deemed necessary to carry on with subsequent steps, or when trying to reproduce a yield claimed from a literature example of a similar or perhaps even the same transformation on the same compound.

The factors affecting real and reported literature yields have been discussed (Wernerova and Hudlicky, 2010). Yields >95% are very rarely found in the publication *Organic Syntheses*, in which articles provide detailed descriptions of the preparation of a single compound, and the procedure has been checked for accuracy in the laboratory of another synthesis practitioner (see Chapter 2, Section 2.3 and Chapter 17). This is in contrast to primary literature that may include experimental procedures and is peer reviewed, but is not actually checked experimentally. The obvious suggestion is that the primary literature is overoptimistic when it comes to reported yields. One possible explanation is that the sample was not a homogeneous species, because the small quantity of product did not permit full purification or accurate demonstration of its purity. Hudlicky also showed that each manipulation such as extraction, filtration, and column chromatography has a 1–2% loss of mass, raising skepticism about any reaction that involves these processes and claims a yield >95%. Purification methods that can be relied upon to give products of very high purity, distillation, and recrystallization, require quite significant quantities of material to practice, have much higher mass losses, and therefore should be expected to give lower yields. When product masses are below 5 mg, common today with lengthy syntheses and the capability to fully analyze spectroscopically tiny amounts of products, just the weighing errors exceed 10%. A bias also may be introduced by authors reporting the best yield ever obtained in a reaction rather than a range of yields they observed when replicating their optimized procedure. As discussed elsewhere in this book, it is widely regarded as very difficult to reproduce yields claimed in literature reports.

The reporting of yields "based on recovered starting material" (sometimes abbreviated BORSM or BRSM, possibly obscuring what is being reported) has increased greatly in the last two decades. Use of this convention may also be indicated by a statement that starting material was "recycled." These terms pertain to reactions that are difficult to carry to completion. One reason for this observation could be that with stoichiometric ratios of reagents to reactant, the reaction slows after an appreciable fraction of reaction has occurred, since concentrations have dropped significantly. If the reaction is stopped at this stage, the remaining reactant may be recovered and the amount that was consumed can be calculated. If every molecule consumed had been converted to product, the BRSM convention would report a 100% yield. The recovered starting material could presumably be resubmitted to reaction conditions to gain more product.

However, a yield quoted on a BRSM basis does not give the chemist 100% of the amount of product calculated as the theoretical yield to take into the next step. To achieve compound throughput equivalent to a reaction that does go to completion, the reaction must be repeated but the results will still fall short. A reaction that reaches 66% conversion on the first try still gives only 88% conversion in total after one repeat. It becomes impractical to "recycle" starting material when conversions are much lower than this. Furthermore, to recycle starting material requires a separation of starting material and product, commonly by chromatography. While chromatography is a frequently used purification method in the research lab, it is expensive and significantly decreases efficiency as a unit operation in any preparative synthesis.

References

Allesø, M., Rantanen, J., Aaltonen, J., Cornett, C., van den Berg, F., 2008. Solvent subset selection for polymorph screening. J. Chemometrics 22, 621–631. http://dx.doi.org/10.1002/cem.1107.

Dubant, S., Mathews, B., 2009. Enabling facile, rapid and successful chromatographic flash purification. Chromatogr. Today 10–12.

Fair, J.D., Kormos, C.M., 2008. Flash column chromatograms estimated from thin-layer chromatography data. J. Chromatogr. A 1211, 49–54. http://dx.doi.org/10.1016/j.chroma.2008.09.085.

Kremer, K.A.M., Helquist, P., 1984. Purification of air-sensitive organometallic compounds by a modified flash chromatography procedure. Organometallics 3, 1743–1745. http://dx.doi.org/10.1021/om00089a025.

Mander, L.N., Williams, C.M., 2016. Chromatography with silver nitrate: part 2. Tetrahedron 72, 1133–1150. http://dx.doi.org/10.1016/j.tet.2016.01.004.

Pirrung, M.C., Deng, L., Li, Z., Park, K., 2002. Synthesis of 2,5-dihydroxy-3-(indol-3-yl)-benzoquinones by acid-catalyzed condensation. J. Org. Chem. 67, 8374–8388. http://dx.doi.org/10.1021/jo0204597.

Ratnayake, W.M.N., 2004. Overview of methods for the determination of trans fatty acids by gas chromatography, silver-ion thin-layer chromatography, silver-ion liquid chromatography, and gas chromatography/mass spectrometry. J. AOAC Int. 87, 523–539.

Still, W.C., Kahn, M., Mitra, A., 1978. Rapid chromatographic technique for preparative separations with moderate resolution. J. Org. Chem. 43, 2923–2925. http://dx.doi.org/10.1021/jo00408a041.

Wernerova, M., Hudlicky, T., 2010. On the practical limits of determining isolated product yields and ratios of stereoisomers: reflections, analysis, and redemption. Synlett 2701–2707. http://dx.doi.org/10.1055/s-0030-1259018.

Williams, C.M., Mander, L.N., 2001. Chromatography with silver nitrate. Tetrahedron 57, 425–447. http://dx.doi.org/10.1016/S0040-4020(00)00927-3.

Yamamoto, Y., Nishimura, K., Kiriyama, N., 1976. Studies on the metabolic products of *Aspergillus terreus*. I. Metabolites of the strain IFO 6123. Chem. Pharm. Bull. 24, 1853–1859. http://dx.doi.org/10.1248/cpb.24.1853.

Methods for Structure Elucidation

15

Chapter Outline

Many useful and detailed texts are available concerning the elucidation of organic structures using spectroscopic and other types of data. This chapter aims to address primarily the practical issues in collecting the data.

15.1 Nuclear Magnetic Resonance Spectroscopy

Proton nuclear magnetic resonance (NMR) requires a nonprotonic solvent, generally deuterated, to avoid obscuring signals of the desired compound by solvent signals. In circumstances where protons from the solvent are unavoidable, special spectroscopic techniques are available to suppress their signals. An internal chemical shift standard, such as TMS (tetramethylsilane), may optionally be included in the solvent. The cheapest organic NMR solvent is $CDCl_3$, because it can be prepared by base-catalyzed exchange in D_2O (Eq. 15.1). It is never 100 atom% deuterium, however, so a small singlet from residual protons in the solvent (i.e., $CHCl_3$) is often seen near 7 ppm. Some workers use this signal as the internal standard. It should also be kept in mind that an intrinsic property of chloroform is that it undergoes slow decomposition to HCl, which will also occur (to DCl) in $CDCl_3$. If sensitivity to acid of the compound under analysis is a concern, another NMR solvent should be chosen. Another commonly used solvent is d_6-acetone, which like chloroform is volatile and permits the easy recovery of the compound from the NMR sample. Other solvents that are often used for polar or ionic species include d_6-DMSO and D_2O, which are not easily evaporated. Essentially any other solvent is available in deuterated form, the main issue being the expense.

Handbook of Synthetic Organic Chemistry. http://dx.doi.org/10.1016/B978-0-12-809504-1.00015-7

$$\text{CHCl}_3 + \text{D}_2\text{O} \xrightarrow{\ominus\text{OD}} \text{CDCl}_3 + \text{HOD} \tag{15.1}$$

When using a Fourier transform NMR instrument, 1 mg of a compound of molecular weight less than 500 Da is sufficient to obtain a quite reasonable proton NMR spectrum in a brief time. It is also increasingly common that such instruments are equipped with a cryoprobe, where the radiofrequency probe electronics are cooled to liquid helium temperatures. Sensitivity is enhanced up to 4× and much less compound can be used. With a cryoprobe, even powerful but demanding two-dimensional NMR experiments (COSY, NOESY, etc.) can be recorded on 1 mg samples. Carbon NMR intrinsically requires a larger quantity of compound than proton NMR because of the low natural abundance (1.1%) of ^{13}C. However, a cryoprobe enhances sensitivity for carbon as well and enables quality carbon spectra to be obtained on even the small quantities of compound being prepared in long synthesis schemes.

To prepare an NMR sample, dissolve the compound in 1 mL of the deuterated solvent, which should be enough to fill the NMR tube to about 5 cm depth. If a smaller volume is used, it is essential to a create a cylinder of constant magnetic susceptibility within the RF coils of the probe. This is accomplished with a susceptibility plug set (Fig. 15.1) or a Teflon vortex plug. The vortex plug comes with a matching threaded metal rod. The plug is gently pushed into the tube with the rod until the bubble of air above the solution disappears. The tube is then capped. After the spectrum is collected, the plug is removed by threading the rod into the plug and pulling, very gently.

It is important to remove solids or particulate matter from the sample, which will affect the magnetic field homogeneity and reduce the resolution of the spectrum. The sample can be filtered through a glass wool plug in a Pasteur pipette, with the filtrate flowing directly into the tube. Dissolved metallic and paramagnetic impurities (e.g., Fe^{3+}, Mn^{2+}) are another factor that can cause reduced resolution. It is difficult to remove such species at sample preparation time. All NMR tubes must be sealed with plastic caps. Color coding based on the cap will help keep track of samples when taking several spectra at one sitting. Some workers use paper tags looped onto their tubes with string for this purpose.

An NMR tube adequate for routine spectra is the Wilmad 6 PP (5 mm) tube. For important, high-quality spectra, the Wilmad 7 PP tube can be used. Chipped, cracked, or broken tubes should not be used, but they might be resurrected by sanding off the broken part using a glass sanding wheel, if the glass shop has one. Tubes may be cleaned by rinsing with ether and acetone. It is important not to use any kind of pipe cleaner or brush to clean the inside of an NMR tube. Scratches will lead to "blips" of sample outside the cylinder defined by the rest of the tube, causing magnetic field

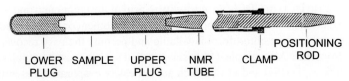

Figure 15.1 A Doty susceptibility matching plug set.
Provided by Doty Scientific.

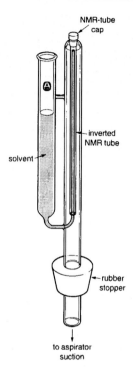

NMR-tube
cap

inverted
NMR tube

solvent

rubber
stopper

to aspirator
suction

Figure 15.2 An apparatus to wash nuclear magnetic resonance tubes with solvents.
© Sigma–Aldrich Co. LLC.

inhomogeneity. KOH baths are not good for NMR tubes, and they should never be placed in any type of chromic acid, as even tiny deposits of paramagnetic metals will severely affect resolution. Aqueous solutions and salts are removed from NMR tubes with much difficulty because of their long, narrow internal dimensions and the surface tension and viscosity of aqueous solvents. A special apparatus is available that enables solvents to be shot to the bottom of an inverted tube and then drain (Fig. 15.2). Some workers dry NMR tubes in an oven, but it is best not to exceed 60°C, meaning that the same oven used for drying glassware cannot typically be used. This precaution is necessary because the thin glass may be distorted by heat. It can introduce curvature relative to the tube's long axis, causing the sample to wobble when spinning, break in the probe, or potentially damage the probe's radiofrequency coil by contact. Tube distortion is also the major source of spinning side bands and increased shimming times.

When a compound has hydrogens bound to O, S, or N, their chemical shift and line shape can be quite variable, depending on parameters such as solvent, temperature, concentration, and impurities. It can be very useful to know which signals in the ^{1}H NMR spectrum correspond to those protons, and this can be determined by showing which signals are "exchangeable"—that is, which can move from molecule to molecule according to the equilibrium in Eq. (15.2). In this example, hydrogens are switched between the "blue" and the "red" molecules, which leads to a phenomenon called exchange broadening. The theoretical treatment of exchange is left to sophisticated NMR spectroscopy texts, but it has a practical impact on the appearance of the spectrum. An obvious way to examine exchange would be to add a deuterated, exchangeable molecule such as D_2O to the sample and observe the disappearance of the signal for the X–H group as it is converted to the X–D group (Eq. 15.3). Typically the spectrum is acquired normally, the D_2O is added, and the spectrum is acquired again for comparison. Because D_2O is immiscible with most organic NMR solvents, vigorous shaking is required for mixing, and of course the heterogeneous mixture may affect magnetic field homogeneity and therefore spectral resolution. Alternatively, MeOD could be used. Another approach is to add a drop of formic acid (miscible with organic solvents) to the sample following acquisition of the first spectrum. Its protons are far downfield and therefore will not interfere with most signals in the spectrum. Its effect is to accelerate exchange (Eq. 15.4), which should change the line shape of the exchangeable protons, and often shifts them downfield.

$$CH_3OH + CH_3OH \rightleftharpoons CH_3OH + CH_3OH \qquad (15.2)$$

$$CH_3OH + D_2O \rightleftharpoons CH_3OD + HOD \qquad (15.3)$$

$$CH_3OH + CH_3OH \underset{}{\overset{HCO_2H}{\rightleftharpoons}} CH_3OH + CH_3OH \qquad (15.4)$$

It should also be emphasized that many NMR facilities have their own policies and recommendations for sample preparation and spectrum acquisition that may preempt the more general advice given here.

The processing of NMR data uses sophisticated software that also converts the graphical spectrum into tabulated spectral measurements ("peak pick") such as integrated area in a chemical shift range, chemical shift in ppm, and chemical shift in Hz that can be used to derive coupling constants. This processing is often done on the computer operating the spectrometer, but can also be done off-line. Workstations may be available outside the spectroscopy lab for this data analysis, but software for this task is also available that runs on modern personal computers. Some is provided by spectrometer vendors, some has been developed by NMR practitioners. Packages are available for Windows, MacOSX, and Linux operating systems, such as mNova, iNMR, and NMRPipe. It has even become possible to do a significant amount of NMR spectral manipulation on tablets or smart phones (Cobas et al., 2015).

Given the essentially limitless digital storage available in 2016, it is reasonable to save all NMR spectra in electronic form (as well as prepare printed copies for ease of use around the laboratory and for archival purposes). Having two electronic forms for each spectrum is best. The raw NMR data, or free induction decays, are one such form. Different Fourier transform/signal processing procedures can be applied to these data at any time to obtain a spectral presentation that differs as needed from what was obtained originally. Another form is the final graphical spectrum. It includes software-added annotations such as peak picks and integrations and should be saved in an image format, such as a TIF or PDF. Many journals *require* the images of the NMR spectra of all new compounds either as part of the evaluation of a paper for publication or as part of its supplementary material. Electronic laboratory notebook systems may also permit the import of such spectral images in digital form.

Comparing the spectrum the chemist has taken today with a tabulated spectrum of the same compound from the literature or an actual spectrum from the several available spectral databases has its pitfalls. Chemical shifts can be quite solvent dependent, so if one is not using the same solvent in which the earlier spectrum was recorded, correspondence of peaks may be imperfect. Chemical shifts are also concentration dependent in some cases. Synthetic chemists almost never concern themselves with concentration in preparing an NMR sample, so this factor typically cannot be taken into account. NMR spectrum prediction software has become available even for personal computers. While this tool can certainly be helpful, the algorithms used in these programs may not be perfect, and there is no substitute for an actual NMR spectrum for visual comparison.

The best source for a comparison spectrum is one that has been taken on the same instrument, which is available if NMR has been used to test reagent purity. To facilitate the direct comparison of starting material spectra with those of reaction products, the chemist should develop the habit of using the same plotting procedure for each spectrum. That is, always plot the proton spectrum between 0 and 8 ppm (or 10 ppm if aldehydes are frequently used) or the carbon spectrum between 0 and 200 ppm on a single sheet. With high-field NMR spectrometers that spread the proton chemical shift range over thousands of Hertz, these plots do lose some visual detail, but this information can be recaptured in subsequent "blow-up" plots (small portions of the spectrum printed at a larger size) of specific regions of interest.

When examining NMR spectra of crude reaction mixtures, extraneous peaks may frequently appear owing to solvents remaining from the reaction or the purification or from contaminants found within them (e.g., dioctyl phthalate or 2,6-di-*tert*-butyl-4-methylphenol). To facilitate identification of such peaks in high resolution NMR spectra, a comprehensive compilation was made by Gottlieb et al. (1997) and Fulmer et al. (2010). These data are tabulated in Appendix 7.

15.2 Infrared Spectroscopy

There is a fair amount of variation in acquiring infrared spectra based on the particular spectrometer used. The reader is referred to the many textbooks, user manuals, and local instrumentation facility guides for further details. Infrared spectroscopy has fallen into disuse in some laboratories, perhaps because of the amazing advances in and immense power of NMR. However, there are still questions about organic compounds that are best asked and answered with IR. Many research advisors have been frustrated by a coworker's conclusion that a compound has no carbonyl group based on the absence of protons in the 2–3 ppm region, rather than using direct information about a carbonyl that comes from the IR spectrum. It is by far the best way to make inferences concerning functional groups present in the molecule.

For liquids, IR spectra are typically obtained using thin films formed between two NaCl windows (also called salt plates). Free-flowing liquids can be dotted neat, using a Pasteur pipette, onto one salt plate as a single drop of as little as 1 mg; holding the other plate by its edge, the first plate is covered with it and they are pressed together, spreading the sample. Provided the two plates have smooth surfaces, a film should be readily seen between them. This assembled sandwich is inserted into a holder for acquisition of the spectrum. If the sample is a viscous liquid or a solid, it is sometimes possible to form a thin film by transferring onto the plate a solution of the compound in a minimum volume of a volatile solvent (diethyl ether, dichloromethane). Evaporation of the solvent under a stream of nitrogen generates the thin film. After acquisition of the spectrum, the sample may be recovered by rinsing it off the plate with diethyl ether, and the plate cleaned with acetone. NaCl windows will inevitably become cloudy and their surfaces will roughen over time. They can be polished by dropping 50% aqueous ethanol onto a paper towel and rubbing the window on the towel in a circular motion, first on one side and then on the other. This wet solution should be removed with an acetone rinse and the plates dried under a stream of nitrogen. Storage in a desiccator is recommended.

While more laborious, infrared spectra can also be obtained in solution, usually in chloroform. $CHCl_3$ has a simple IR spectrum that does not interfere with most signals in the spectrum of the sample, and a reference cell can be used to eliminate those. A solution spectrum of a liquid sample may be obtained to examine bands whose appearance is concentration dependent, for example OH stretches. A solution spectrum of a solid sample may be obtained because the thin film technique described earlier failed. IR solution cells have two salt windows held a fixed distance apart, with the space between accessed through filling ports (usually Luer connectors) that can be plugged. Approximately 3 mg of the sample is placed in a vial and dissolved in 4 mL of solvent. The solution is taken up in a 1-mL syringe and injected into one of the two open ports of a solution cell. One plug is inserted, the cell is tilted to allow air to escape, and the other plug is inserted. If available, a reference cell is filled with pure solvent in the same way. Sample recovery can be accomplished by flushing the solution out of the cell into a beaker with a nitrogen stream. Several rinses of the cell with dry solvent followed by nitrogen flushing are recommended. NaCl solution cells are much more expensive than salt plates and must be stored in a desiccator. Many chemists also acquire IR spectra of solids in KBr pellets, whose preparation and use will be left to other, older, texts.

The foregoing methods are being used less as IR spectrometers with universal attenuated total reflectance accessories become more common. Here, the solid or liquid sample can be deposited directly onto a diamond window for spectral acquisition. As there is effectively no sample preparation with this method, there should be no barriers to the use of IR for compound characterization.

15.3   Ultraviolet Spectroscopy

Using volumetric techniques, a solution of precise molar concentration is made up in the desired spectroscopic grade solvent. Common solvents include diethyl ether, ethanol, hexane, and cyclohexane. The concentration is chosen such that the combination of the path length of the cuvette that will be used and an estimate of the extinction coefficient based on literature values will produce an absorption greater than 1. Calculation of the estimated absorption is based on the Beer–Lambert law (Eq. 15.5), where c, concentration in M, d, path length in cm, and ε, extinction coefficient in M^{-1}cm^{-1}.

$$A = \varepsilon cd \tag{15.5}$$

Cuvettes for UV absorption spectroscopy (Fig. 15.3) are made from quartz, which absorbs very little light in the 200–400 nm range. This is in contrast with conventional borosilicate glass (Pyrex), which absorbs significantly at $\lambda < 300$ nm. Generally, UV cuvettes come in pairs with matched optical properties. Quartz inserts are also available to reduce the path length of light through the solution from 1 to 0.1 cm or even 0.01 cm. When a compound has both strong and weak absorption bands, which is common with carbonyl groups, such inserts can permit the UV spectra of both bands to be obtained from a single solution. The solution is placed in a sample cuvette, and pure solvent is placed in a reference cuvette. The optical surfaces are wiped clean with tissue and the spectrum recorded. After solvent removal, the cuvettes are rinsed with pure solvent and flushed dry with nitrogen.

15.4 Combustion Analysis

This technique simply provides the percentage of designated elements (at a minimum, C and H) in a bulk sample, and from these data an empirical formula can be

Figure 15.3 A quartz UV cuvette with Teflon cap.

derived (assuming the sample is a single molecular entity). Comparing the observed percentages to those calculated from the proposed formula enables it to be validated or refuted. Naturally, there is error in the analytical method; the permitted difference for each element between the calculated and observed percentages is 0.4%. This 0.4% difference is measured in absolute terms, i.e., if the calculated percentage of H in a formula is 7.44%, then analytical results between 7.84% and 7.04% are considered acceptable.

If a compound is a solid, it should be recrystallized before combustion analysis, preferably from a volatile solvent. It should then be pumped on overnight in the vial in which it will be sent for analysis. If it is a liquid, it can be purified by chromatography or distilled. Kugelrohr distillation is often used for this purpose—recall that it merely separates volatiles from nonvolatiles, but it is often effective in removing the slight amount of color in a compound that should have none. If the compound has been purified by chromatography, is a single spot on TLC, is pumped on overnight, and is pure, it will frequently "hit" an acceptable elemental percentage. If it is a viscous liquid, removing all of the solvent may be more difficult. Heating with a heat gun may be required. A somewhat archaic piece of glassware to warm a sample under vacuum, a "drying pistol" (Fig. 15.4), can also be used.

Some compound classes, such as nucleosides, are highly hygroscopic and analyze as hydrates. Recalculating the percentages of each element to include some water is acceptable with such difficult compounds. However, the only acceptable stoichiometries are 1 H_2O and ½ H_2O. Adding in fractional molecularities of water to get the combustion analysis to hit is manipulating the data. The amount of compound required for a combustion analysis is set by the policies of the analyst, but typically is 10 mg for carbon, hydrogen, and, if the compound includes it, nitrogen (CHN) in duplicate. More sample is required if other elements are to be determined.

15.5 Mass Spectrometry

Mass spectral fragmentation patterns have received strong emphasis in older teachings of organic structure elucidation. However, these data are most available when using electron impact for ionization, and this ionization method is falling into disuse. More modern ionization methods are "soft," meaning that they do not lead to fragmentation. The molecular ion is therefore detected, which after all is the prime piece of information sought in a mass spectrum. Soft ionization methods include fast atom bombardment, matrix-assisted laser desorption ionization, electrospray ionization, and chemical ionization, particularly atmospheric pressure chemical ionization. These methods typically create charged adducts of analyte molecules with H^+, metal ions, or both. The loss of the information that was formerly available from fragmentation patterns is a small sacrifice, as information on molecular substructures is now more reliably determined from modern two-dimensional NMR techniques.

For the most part, obtaining a mass spectrum of a synthetic product has meant submitting a sample to a service facility. This is changing, though, as many facilities are taking advantage of "open access" LCMS instruments to permit chemists to

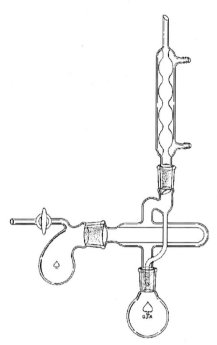

Figure 15.4 The Aberhalden drying pistol. A solvent with a boiling point below the melting point of the solid to be dried is chosen. Its refluxing around the barrel of the vacuum chamber warms the solid to drive off any traces of solvent. If water is to be removed, a desiccant can be placed in the reservoir to the left (in the "handle").

obtain MS data themselves. Sample preparation protocols are unique to each type of instrument and facility, as is the software that analyzes the data. A capability often available in software, however, is matching of the molecular ion (accounting for the fractional masses of the elements) and the pattern of isotope peaks to possible molecular formulas.

15.6   Optical Rotation

The rotation of plane-polarized light was the first measurement chemists made involving chirality. As a molecular characterization technique, a significant problem with rotation is that it is a bulk property representing the sum of the rotatory power of all compounds in the solution. This means that a small quantity of a highly rotating impurity can swamp a major compound that has a small rotation. Consequently, it is crucial that molecules whose optical rotation is measured be fully purified to a high level (ideally >99%) and analyzed to establish their purity.

   Optical rotation is dependent on compound concentration but is not perfectly linear with it, as is typical for other optical measurements. Therefore, it is important that concentration be accurately determined and specified in the data. For most compound characterizations, as small a sample as is practical is used. For spectroscopic techniques, milligram quantities often suffice and concentration is immaterial. However, to accurately weigh a sample for a rotation, even with an analytical balance, >10 mg should be used. With volatile solvents, typically alcohols or dioxane, the sample can be recovered of course. As in recording a UV–Vis spectrum, spectroscopic grade solvents should be used. Chloroform has seen much use for rotations and may be used to compare current data to older literature, but is disfavored for new characterizations. It has several drawbacks, including its toxicity (Appendix 5). It must be purified (filtration through basic alumina at least) because on storage it spontaneously decomposes to a variety of products including HCl. It also is important to note what stabilizers, if any, are present in the chloroform. Ethanol or amylenes are sometimes added to commercial samples of chloroform to inhibit decomposition, and their presence can affect measured rotations. Rotations are dependent on temperature, which is why polarimeter cells have a thermostatic jacket. Given all these caveats, it seems that rotations are capricious and less than reliable for comparing the enantiomeric constitution of different samples. While true, rotation is also the only technique for measuring optical activity that can be performed in most laboratories and gives easily tabulated data that has archival value because it is independent of the specifics of the method (unlike chiral chromatography, in the next section).

A sample solution of 2 mL should be prepared in a volumetric flask. This volume should be sufficient to measure a rotation in a commonly used 1 mL polarimeter microcell with a path length of 1 dm. A decimeter (dm) is a rarely used unit and of course is a one-tenth of a meter, or 10 cm. It is crucial the solution includes no particulate matter and that air bubbles be eliminated, as these can scatter light and corrupt the measurement. Filtering the solution before taking the rotation is advised. The cell is filled via a syringe or pipet, mounted on the instrument, and the wavelength is selected. Often simply the yellow-colored sodium D line (at 589.3 nm) is used. If a molecule has low or no rotation at the D line, other wavelengths can be tried to find one with a greater rotation. The rotation measurement itself should have many replicates; their average is used for the calculation of specific rotation. It is important to thoroughly clean the polarimeter cell afterward to remove all traces of chiral compound. Cells are expensive and therefore reused many times, and any carryover to future samples will corrupt their rotation data.

One caution concerns molecules with large rotations. If the rotation is larger than 180 degrees, a single measurement cannot reveal this. It also cannot distinguish +270 and −90 degrees, for example. If such behavior is suspected, a cell with a shorter path length can be used or the solution can be diluted for a repeat measurement. If the rotation decreases by the same factor that the amount of compound in the polarized light beam has been reduced, the measurement is well behaved and either rotation measurement should be valid for determining the specific rotation. If the rotation changes in some way other than this, the first rotation value should be discarded and the specific rotation calculation based on the sample with the lower quantity of compound. In an abundance of caution, record the rotation of *every* sample at two different concentrations, which should not be related by a factor that is an integer.

Specific optical rotation is reported in the form of Eq. (15.6). As superscript and subscript to the bracketed α, the temperature (°C) and wavelength at which the rotation was taken are specified. If the sodium D line was used, only a D is needed. Otherwise, a wavelength in nm can be specified. If the rotation is performed at 20°C, the temperature need not be listed. The specific rotation calculated as described later is listed next, to no more than two decimal places. In parentheses is given the concentration of compound in the unusual units of g/100 mL, again to no more than two decimal places. Because these units are not used much by chemists, it is easy to make a mistake in the calculation and report concentration incorrectly. The true units of specific rotation are also unusual (degree mL/g dm) but mostly rotation is reported simply in degrees and the other elements of the unit are understood to be present but not listed.

The calculation of specific rotation is performed using Eq. (15.7). The measured rotation (in degrees positive or negative) is read from the polarimeter. The factor of 100 takes into account the concentration units (g/100 mL). The path length of the cell l is in decimeters and the concentration c is in g/100 mL. For the microcell mentioned earlier, the path length is 1 dm.

$$[\alpha]_D^{25} = \text{-8.0 } (c\ 0.2,\ \text{MeOH}) \tag{15.6}$$

$$[\alpha] = \frac{100 \cdot \text{measured rotation}}{l \cdot c} \tag{15.7}$$

Besides the use of rotation data for documentary purposes, they are used to measure the ratio of enantiomers in a particular sample (subject to the earlier caveats about rotations). The enantiomer ratio is a crucial measure for many modern synthetic methodologies that have as their goal the creation of single enantiomer products. Samples of chiral compounds can be classified as racemic, scalemic, or holemic based on their enantiomeric composition. Racemic means a 50:50 mixture of the two enantiomers, scalemic means an enantiomer ratio other than 50:50, and holemic means only a single enantiomer. A racemic material has an optical rotation of zero, of course. A holemic sample has the maximum rotation that could be observed for that compound under a specified set of conditions, which may have been reported. A scalemic material should fall somewhere in between. If the current sample of compound has a rotation that matches the past report for a holemic sample, this is evidence it is also a single enantiomer. If the rotation is less than the reported maximum rotation, that suggests both enantiomers are present in unequal amounts, so the sample is scalemic.

The specific rotation observed for the current sample can be compared to the maximum rotation to determine the proportion of enantiomers. For example, if the specific rotation measured is 90% of the maximum, the current sample must be 95% one enantiomer and 5% the other enantiomer. This could be reported as an enantiomer ratio, sometimes abbreviated as *er*, of 95:5. In older literature, the measurement might be reported as enantiomeric excess (the percentage by which one enantiomer exceeds the other), abbreviated *ee*. The *ee* of the same sample would be 90%. Put another way, it is 90% one enantiomer and 10% racemic mixture. One virtue of *ee* is that its value is the same as the percentage of the maximum rotation shown by the compound. However, *ee* requires the mental gymnastics to convert 90% to a 95:5 ratio, a more explicit way of understanding molecular mixtures that chemists often prefer, so the use of *ee* is being discouraged.

The enantiomeric ratio can be determined using other techniques, such as chiral chromatography (discussed in the following section), NMR chiral shift reagents, and chiral derivatizing agents. Whenever possible it is wise to verify the enantiomeric ratio determined by optical rotation using one of these other methods at least once, to demonstrate the rotation is well behaved for the specific compound under study.

15.7   Chiral Chromatography

This method uses all the elements of a standard HPLC system (Section 10.3) with the exception of the column, which includes a chiral stationary phase (CSP). The two enantiomers of a compound interact in diastereomeric ways with the single enantiomer of a chiral selector on the stationary phase. Therefore, the affinity for the CSP of each enantiomer is different and their chromatographic mobilities are different. The difference of retention times for the enantiomers gives two peaks on the chromatogram that can be digitally integrated. A very accurate measurement of the ratio of enantiomers results. Chiral chromatography is done in both normal phase and reverse phase.

Identifying a CSP to determine the enantiomeric ratio of a sample is straightforward in principle—the racemic compound is first analyzed, with the goal being a chromatogram showing two peaks of equal area fully separated at the baseline. However, it is not possible to predict based on structure which CSP will resolve or separate a particular racemate, so empirical testing of different CSPs is required. Vendors of CSPs even offer screening of CSP columns as a service.

Chiral chromatography can also be used preparatively to obtain samples of pure enantiomers from a racemic mixture, provided it is done on the relatively small scale compatible with HPLC. This is commonly done in early stages of the development of chiral drug candidates to identify the biological activity of the individual enantiomers. Everything that has been described earlier for HPLC is also true for GC using a CSP, with the exception of the ability to perform preparative separations.

15.8   Crystal Growth for X-Ray

It is increasingly common for synthetic chemists to use X-ray crystallography to prove the structures of intermediates, and even to solve the structures themselves. Powerful computer software has put this task within the reach of many graduate students. However, growing of the single crystals needed to obtain an X-ray structure is still a bit of an art, and determination of when a crystal is likely to give a good data set requires experience. If a crystallographer is nearby, they are a wealth of assistance and information.

In the compound purification Section (14.9), techniques for crystallization are described. Growing single crystals suitable for X-ray diffraction is a bit different. The main point is that a crystal of sufficient size is needed. A crystal should be at least 0.2 mm in at least two dimensions for best results, which is assessed using a microscope. Crystals grow from nucleation sites, so having as few of such sites as possible is desirable to make crystals larger. Sites of nucleation include microscopic particles and defects, so the cleanliness of the glassware and solvents used for crystallization is crucial. Mechanical agitation can fracture crystal seeds or promote new nucleation sites, so it is important to isolate the crystallization from all sources of vibration. Likewise, the chemist should leave the crystallization alone as it proceeds— ideally, up to a week without any contact. Rapid crystallization can make formation of additional nucleation sites competitive with growth from seeds, whereas slow crystallization encourages growth of existing crystals.

Observing the foregoing, the main goal in crystallization is to begin with a solute that is very near its solubility limit, then slowly reduce the solubility. This can be done by evaporation of the solvent, but the more volatile solvents evaporate too quickly for this to work well. Slow cooling works similarly to reduce solubility but is easier operationally. It is performed with a well-insulated Dewar filled with hot water into which a test tube of the crystallization mixture is inserted. Another method called vapor diffusion does not use temperature changes but rather requires a crystallization system based on two solvents, similar to a situation described in Section 14.9. Ideally, the solvent in which the compound is less soluble (=antisolvent) is also somewhat volatile. The solute is placed in a test tube in the solvent in which it is more soluble. The test tube is placed in a beaker that also holds the antisolvent (Fig. 15.5), and the whole apparatus is sealed with parafilm, plastic wrap, or aluminum foil. Over time, exchange of liquid between the pure solvent in the tube and the antisolvent in the beaker occurs via the vapor phase. This process increases the proportion of antisolvent in the tube, reducing the solubility of the solute and inducing crystal growth.

15.9 Novel Compound Characterization

To prove beyond doubt that one has prepared a novel structure that is presented in a peer-reviewed publication, several types of data are expected. Innumerable papers have been delayed simply because all of the required data had not been obtained when

Figure 15.5 Vapor diffusion set-up for growth of single crystals.

the compound was first prepared, and someone had to go back to get it. Needed spectroscopic data include ^1H NMR and ^{13}C NMR, plus spectra of other nuclei if present, most often ^{19}F or ^{31}P. These techniques are qualifying or disqualifying about the structure as a whole—one often hears an NMR spectrum "proves" a structure, though this is an overstatement—a preponderance of evidence establishes structure. An IR spectrum is required, and if the molecule has a significant chromophore, a UV spectrum should also be reported. Though these latter two techniques do not give complete structural information, they are qualifying or indicative for specific functional groups or structural subunits.

Evidence supporting the proposed molecular formula must also be presented. There are two choices here: the traditional combustion analysis, and the more modern high resolution mass spectrometry (HRMS). While the information drawn from a combustion analysis is strictly speaking an empirical formula, not a molecular formula, there really is no ambiguity there since its mode of synthesis is known. A dividend provided by combustion analysis of a new compound is that a successful one shows it was pure. If it was contaminated by any other compound or solvent with a significantly different empirical formula, the observed percentage of each element would not be within the experimental error of the analysis. This can mean that a combustion analysis not within acceptable variance is evidence of a lack of purity, not of an erroneous structure. Repeated cycles of compound purification and resubmission for combustion analysis are often necessary until one finally hits.

The laborious task of combustion analyses makes MS methods attractive, which use the fractional mass (to 0.0001 Da) of the molecular ion to calculate the molecular formula based on the fractional masses of the major isotopes of its constituent elements. With modern soft ionization techniques, it is highly probable that a molecular ion can be observed. The one thing HRMS does not provide is the evidence of purity. In fact, the molecule could even be a small component of a bad mixture, yet still the molecular ion from that particular formula could be detected. As a consequence of this situation, most journals require independent evidence that the compound was purified to homogeneity if HRMS is used to provide the molecular formula. The most common data used for this purpose are an analytical HPLC chromatogram or the actual ^1H NMR spectrum (not just the tabulation of it), showing no extraneous peaks anywhere.

Other required characterization data may not provide direct structural information but can be helpful to workers who later aim to prepare the same compound and need measures to verify its identity. Others may simply need the data for their own purposes. If a compound is chiral and nonracemic, its optical rotation should be reported. Physical state—i.e., amorphous solid, oil, cubic crystals, needles, etc.—and color should be as well. The same applies to the melting point, which should be obtained on a sample that has been recrystallized, and the solvent used for crystallization should be indicated. A chronic problem with novice writers of experimental descriptions is that solid compounds are obtained but they are not recrystallized and no melting point is obtained.

Many of these dictums about compound characterization can also be found in the instructions to authors for a respected journal, such as *J. Am. Chem. Soc.* or *J. Org. Chem.* These and other organic chemistry journals use automated tools to check the experimental data provided in full papers for accuracy.

Conventions should be followed in the presentation of NMR data in an experimental description of a new compound. Those discussed come specifically from American Chemical Society journals, but other respected publishers offer similar guides. Give ^{1}H NMR chemical shifts to two decimal places, followed by parenthesized information on the number of protons in the signal, multiplicity, and coupling constant (J). The last item may be reported to one place after the decimal, but this is not always a significant figure even though it can be read from a peak pick tabulation, as discussed further. Give ^{13}C NMR chemical shifts to one decimal place; be consistent in the listing order of chemical shifts; i.e., from high to low or low to high throughout the experimental section.

It is often found that software-assigned peak frequencies give slightly different coupling constants (especially digits past the decimal) to signals that are obviously coupled (e.g., the CH_2 and CH_3 of an ethyl group). Since J coupling must be symmetrical (A couples B with the exact same J as B couples A), this observation points to a lack of significance of the digits past the decimal. If these "raw data" were listed in an NMR tabulation, they could confuse readers about which signals are coupled. Therefore, if the software peak pick results in different values for the J coupling of signals that are known (from independent evidence, including simply the structure) to be coupled, it is best to take the average of the two values and list the same J for both signals.

An ideal graphical NMR spectrum prepared for publication/archival purposes should be scaled such that the tallest peak in the spectrum arises from the compound being studied, allowing any taller solvent peaks to go off scale. The region shown should be at least −1 to 9 ppm for ^{1}H and −10 to 180 ppm for ^{13}C. Especially with higher field ^{1}H spectra, it can be difficult to perceive complex signals at normal size on a single sheet of paper, so those regions should be expanded and shown in insets ("blow-ups") placed on available white space. The spectrum should be annotated with the nucleus being measured and its nominal spectrometer frequency (i.e., the same magnetic field gives a frequency of 300 MHz for ^{1}H and 75 MHz for ^{13}C), the molecular formula of the solvent, and a clear structural drawing. The software should annotate each signal with the integration and chemical shift. The solvent peak or another peak that is used as the internal chemical shift standard should be marked.

References

Cobas, C., Iglesias, I., Seoane, F., 2015. NMR data visualization, processing, and analysis on mobile devices. Magn. Reson. Chem. 53, 558–564. http://dx.doi.org/10.1002/mrc.4234.

Fulmer, G.R., Miller, A.J.M., Sherden, N.H., Gottlieb, H.E., Nudelman, A., Stoltz, B.M., Bercaw, J.E., Goldberg, K.I., 2010. NMR chemical shifts of trace impurities: common laboratory solvents, organics, and gases in deuterated solvents relevant to the organometallic chemist. Organometallics 29, 2176–2179. http://dx.doi.org/10.1021/om100106e.

Gottlieb, H.E., Kotlyar, V., Nudelman, A., 1997. NMR chemical shifts of common laboratory solvents as trace impurities. J. Org. Chem. 62, 7512–7515. http://dx.doi.org/10.1021/jo971176v.

Cleaning Up After the Reaction   16

Chapter Outline

16.1   Waste Disposal

Waste must be properly stored in the laboratory until containers are filled and the hazardous waste operation of the institution removes them for disposal. It is essential to keep a record of all contents of each waste container, and a label or log is attached to it for this purpose. Many workers use emptied reagent bottles to store waste; it is important to completely deface their labels so it is clear they are waste and not the reagent. Secondary containment is a must for hazardous waste storage in the lab prior to pickup. Bottles should be placed in plastic trays to contain any spills that occur in adding waste or mishaps.

An essential feature of storing waste in the lab is the grouping of similar types of waste into aggregate containers. At a minimum, waste will be classified into aqueous, halogenated organic, and nonhalogenated organic types. Waste may also be segregated by hazard class (See Appendix 1), but institutional waste disposal may not accept it directly in those forms, as they include reactive chemicals. In such cases, they must be deactivated appropriately, converting them to waste that falls into one of the three essential classes first mentioned. The greatest quantity of waste generated in a synthetic lab is solvents. Since additions to waste containers are being made often, there is a natural tendency to leave them open, but this violates safety codes. The best solution to this dilemma is a funnel with a hinged lid (Fig. 16.1).

A material that may not usually be thought of as waste but should be treated as such is used vacuum pump oil. Because cold traps are not used diligently and are imperfect, pump oil is likely to be contaminated by any compound with which the pump was used. Pump oil should receive its own designated waste container as well.

Naturally, each organization has its own policies and procedures for hazardous waste disposal that may preempt the more general advice given here.

16.2   Cleaning Equipment

It is common to find ground-glass joints that have become frozen or stuck during a reaction. This is typically the result of a reagent or solvent leaching away the grease. Approaches to freeing stuck joints are presented in Appendix 10.

Handbook of Synthetic Organic Chemistry. http://dx.doi.org/10.1016/B978-0-12-809504-1.00016-9

Figure 16.1 This plastic funnel has a lid that enables easy access to the waste bottle but also meets the requirement that the container be closed when not in use.
Thermo Fisher Scientific Inc.

When glassware has been used for reactions involving bases, and in some other instances such as aluminum or other metal reagents, a fog may be seen on the glass that may be a metal hydroxide. These films hold onto the glass surface tightly, but rinsing the glassware with dilute HCl may remove them. For stubborn organic residues on glassware, including greases, chlorinated hydrocarbons often work best to remove them. When none of these tactics work to clean up glassware, the chemist often considers a bath of chromic acid (or modern and more environmentally friendly cleaners such as Nochromix). These methods may be very useful for volumetric apparatus in analytical chemistry; however, they are much less successful for organic residues. Standard cleaning measures recommended by glass manufacturers include hot water and soap or detergent such as Alconox.

Another choice is a cleaning bath formed from KOH in an alcohol solvent such as isopropanol. This type of cleaning will likely be required by a glassblower before he or she will work on a piece of glassware that has been used for reactions. This treatment is very effective at removing all traces of silicon grease from joints and stopcocks, especially if warmed, but hydrocarbon grease (e.g., Apiezon) is impervious to a base bath. Glassware can be immersed for up to 30 min, but be certain not to leave glass in an alkali bath longer than needed to clean it. Prolonged immersion, even at ambient temperature, damages ground-glass joints, dissolves glass frits, and leaves glass surfaces etched. Remove the item using gloves or tongs, but care must be taken because it will be slippery. Following cleaning in a base bath, a short soaking with 1 N HNO$_3$ will stop the attack of alkali on the glass.

The cleaning solution for this bath is prepared either by dissolving 100 g of KOH pellets in 50 mL of H$_2$O and, after cooling, making up to 1 L with isopropanol or by adding 1–2 L of 95% ethanol to 120 mL of H$_2$O containing 120 g of NaOH or KOH. These baths have a relatively long lifetime, limited only by their effectiveness (in the judgment of the chemist). Only a steel container should be used for such an alkali bath. Glass could be easily shattered, spreading caustic solution everywhere, and of course will eventually be dissolved. Should the worst happen and the alkali/alcohol bath catch fire, plastic containers would melt and spread flaming caustic solution everywhere.

Syringes must be disassembled for cleaning, following which they may be dried in batches, cooled, and stored in a desiccator for future use. Drying is done in an oven (>130°C); usually 30 min are sufficient. Tuberculin syringes are numerically coded on the barrel and plunger and should be reassembled by matching the numbers. If one part of a syringe is broken, many groups maintain a store of parts from which another plunger or barrel may be obtained. If the fit seems good, it is permissible to use syringes with mismatched numbers. If no fit is found, the unbroken part may be left in the store. Multifit syringes do not have numbers, so all fitting must be by trial and error.

Cleaning solutions must be pushed or pulled through most needles. Recall from the discussion of syringe use that reagent remains in the syringe needle following the transfer. This liquid must be removed and/or quenched before cleanup. When using organometallics, aqueous tetrahydrofuran provides a useful quenching agent that will keep the generated metal hydroxide in solution. Mg and Al reagents may leave residues that require dilute HCl to remove. Sometimes needles come with cleaning wires that can be inserted to remove obstructions. The wire can be held with pliers. If needles (or syringes) become drastically clogged and/or stuck, an ultrasonic cleaner will often resurrect them.

The glass frits of filter funnels can easily become clogged and are a challenge to clean. Copper or iron salts can be removed with hot hydrochloric acid plus potassium chlorate. Aluminate and silicate residues can be removed with 2% HF followed by concentrated sulfuric acid. Rinse immediately with distilled water followed by a few milliliters of acetone. Repeat rinsing until acid is undetectable with pH paper.

Specific Example

17

Chapter Outline

This example is one step (boxed) of a multistep sequence (Eq. (17.1)) to prepare a compound useful in synthetic methodology.

$$(17.1)$$

17.1   The Experimental

To 42.4 g (0.5 mol) of 2-hydroxybutyronitrile containing one drop of concentrated HCl, 48 mL (36 g, 0.5 mol) of ethyl vinyl ether was added at a rate such that the temperature was maintained at c. 50°C. After the addition was complete, the mixture was heated at 90°C for 2 h. Distillation directly from the reaction flask provided 65.7 g (84%) of 2-(1-ethoxyethoxy)-butyronitrile as a mixture of diastereomers, bp 84–96°C (18 Torr).[1]

[1] Reprinted with permission from Heathcock, C.H., Buse, C.T., Kleschick, W.A., Pirrung, M.C., Sohn, J.E., Lampe, J., 1980. Acyclic stereoselection. 7. Stereoselective synthesis of 2-alkyl-3-hydroxy carbonyl compounds by aldol condensation. *J. Org. Chem.* **45**, 1066–1081. Copyright 1980 American Chemical Society.

Handbook of Synthetic Organic Chemistry. http://dx.doi.org/10.1016/B978-0-12-809504-1.00017-0

17.2   The *Org. Syn*. Prep

A 1-L, three-necked, round-bottomed flask is equipped with a condenser topped with a calcium chloride drying tube, a magnetic stirring bar, a 500-mL pressure-equalizing addition funnel, and a thermometer. The flask is charged with 174 g (2.05 mol) of 2-hydroxy-butanenitrile to which 0.5 mL of concentrated hydrochloric acid has been added. The addition funnel is charged with 221 g (3.07 mol) of ethyl vinyl ether (Note 1), which is then added dropwise to the stirring cyanohydrin at such a rate that the temperature is maintained at c. 50°C. When the addition is complete, the mixture is heated to 90°C for 4 h. The condenser is replaced with a distillation head and the dropping funnel and thermometer are replaced with stoppers. Direct distillation of the gold-yellow solution from the reaction flask yields 226–277 g (70–86%) of nearly pure 2-[(1′-ethoxy)-1-ethoxy]butanenitrile, bp 85–84°C (30 mm), as a colorless liquid (Note 2).

   Notes.

1. Ethyl vinyl ether was obtained from Aldrich Chemical Company and was used without further purification.
2. The IR spectrum (neat) shows absorption at 2970, 1425, and 1385 cm^{-1}. The C≡N absorption is not observed.[2]

17.3   Comparison

An obvious difference between these two descriptions of the same transformation is that the *Org. Syn.* prep describes the apparatus in some detail. It also is conducted on a much larger scale. Even the experimental preparation is far beyond the 25 mmol scale that has been the focus of this book.

---

[2] Reprinted with permission from Young, S.D., Buse, C.T., Heathcock, C.H., 1985. 2-Methyl-2-(trimethylsiloxy)pentan-3-one. [3-Pentanone, 2-methyl-2-[(trimethylsilyl) oxy]-]. Org. Synth. 63, 79–85. Copyright 1985 Wiley.

Strategies for Reaction Optimization

18

Chapter Outline

When conducting a reaction from the literature for the first time, the outcome (yield, purity, time, and convenience) is unlikely to be as good as reported, or even as good as will likely be the case once the chemist has experience with it. Because the written detail available on a reaction is far exceeded by the amount of expertise and chemical know-how that goes into conducting it, this is to be expected. The chemists reporting the reaction presumably did a significant amount of optimization and provided their "best" procedure. The need for extensive experimental variation to meet or exceed the published results is not expected. However, there are also variations in the source and quality of reagents, available apparatus, and even the climate that can affect reaction outcomes. We experienced this in our lab in the preparation of the very moisture-sensitive nucleoside phosphoramidites (Fig. 14.4), the building blocks of DNA synthesis. This chemistry was developed in the high altitudes and low humidity of Boulder, Colorado, but when conducted in the summer in muggy Durham, North Carolina, reactions often gave only half the reported yields. Although optimization of conditions for known reactions is not necessary, a fresh perspective or different set of chemical experiences or skills can always be brought to bear to improve upon earlier reports. Such contributions should be welcomed by the community of synthetic chemists and are very worthy of publication.

Optimization efforts take on much greater importance when developing new reactions. A cliché of the scientific method is "change only one variable at a time." While good advice to the novice scientist (e.g., one in grade school) involved in deductive investigation, it is a dictum that is widely overapplied. It is much less useful for the inductive investigations of synthetic organic chemistry. The reasons behind this are well-known and readily illustrated with a three-dimensional plot (Fig. 18.1) of the outcome of an experiment (**A**) versus two variables, calcium chloride and magnesium chloride concentration. Assume that the initial experiment was conducted with 150 mM $CaCl_2$, giving **A** = 0.2 (the point nearest the viewer). Changing only one variable at a time means being restricted to movement along the axes of this plot. Adding increasing concentrations of $MgCl_2$ causes a decrease of **A**, so the experimenter would (erroneously) conclude that $MgCl_2$ should not be present. Decreasing the concentration of $CaCl_2$ causes an increase of **A**, until its concentration is very low, when **A**

Handbook of Synthetic Organic Chemistry. http://dx.doi.org/10.1016/B978-0-12-809504-1.00018-2

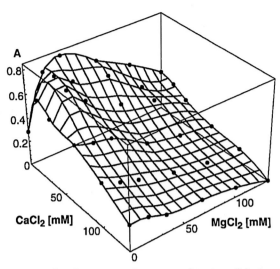

Figure 18.1 The outcome (**A**) of many experiments as a function of the ion concentration. Reprinted from Zauner, K.P., Conrad, M., 2000. Enzymatic pattern processing. Naturwissenschaften 87, 360–362, Figure 1. Copyright 2000, Springer. With permission of Springer.

drops. The conclusion of these experiments, following the one variable at a time dictum, is that the optimum experimental conditions are low $CaCl_2$ and no $MgCl_2$. The appearance of the complete experimental response surface shows that the optimum outcome is with low $MgCl_2$ and no $CaCl_2$, however. The lesson from this study is clear: chemists aiming to optimize novel processes where the important parameters cannot be known in advance should aim to examine variations in a number of experimental variables.

Statistical methods to design and interpret experiments involving multiple variables have been available for some time. They have likely not been applied in synthetic chemistry as broadly as they might be, though texts describing such work are available (Carlson and Carlson, 2005). There are several reasons for this. One is the dogma that only one variable should be changed at a time. Another is the difficulty nonspecialists have in using the computer programs that perform multivariable optimizations. Another is the number of experiments that may be required to adequately map out the experimental response surface (note the number of data points in Fig. 18.1). However, the relatively recent interest in methods for high-throughput and automated experimentation has made collection of these data much less burdensome. Yet it might also be argued that the capability to perform many experiments easily enables optimization to be performed empirically, without the need for statistical design. At the very least chemists could avoid the pitfall described in the example by selecting a number of values for each variable and examining a reasonable number of the combinations that differ from the initial conditions.

Another limitation of statistical design methods is that they apply only to continuous variables, such as concentration and temperature. Experienced synthetic chemists know that nonnumerical variables such as solvent, order of addition, and workup methods can profoundly affect reaction outcome. The ability to conduct many trials of a reaction is often the *only* means to be certain that the best procedure has been found. In the industrial setting, hardware permitting high-throughput experimentation may be available, but of course the chemist still needs to tell a robot what experiments to perform. The laboratory experience gained from optimizing synthetic reactions is indispensable to this decision making.

Strategies to perform many trials of a reaction are not complex. Reaction scale can be kept as small as possible to minimize starting material consumption. At the lower limit (0.5 mg or less), reactions can be designed only for analysis of the outcome by TLC, HPLC, or GC, with no intention of isolating the product. The chemist is looking merely for a single product with properties consistent with the expected compound. Racks of vials can be used as reaction vessels for a dozen or more reactions. In this sort of broad screening, the reactions need not be the same. For example, there are hundreds of reagents to oxidize secondary alcohols to ketones. Replicate vials with the reaction substrate could be set up and a different oxidant added to each. These sorts of approaches are essential to multistep, target-directed syntheses, where precious intermediates at the "frontier" of the total synthesis have been hard won through weeks or even months of effort.

Statistical methods of reaction optimization are commonly applied in industrial research groups doing process development, a subfield of synthetic organic chemistry especially important to the pharmaceutical industry. They are tasked with making drug candidates on substantial scale, much larger than the 25 mmol scale that is the upper limit for this book, to provide supplies for biological testing. Such groups have found shortcuts to conventional statistical design paradigms. Performing a dozen or so variants of a reaction with randomly chosen changes to reaction parameters (e.g., time, temperature, stoichiometry) can be very informative (Hendrix, 1980). Analysis of these trials can predict the likelihood of improving the reaction with further trials, aiding the chemist in deciding if further optimization should be pursued or another synthetic tactic should be examined. The process can be iterated with changes in reaction parameters, accommodating noncontinuous variables such as changing a reagent or solvent.

References

Carlson, R., Carlson, J., 2005. Design and optimization in organic synthesis. In: Data Handling in Science and Technology, vol. 24. Elsevier, New York, pp. 1–574.

Hendrix, C.D., 1980. Through the response surface with test tube and pipe wrench. Chemtech 10, 488–497.

Appendix 1
Safety Protocols

A1.1 Flammable Liquids

A1.1.1 Hazard Overview

A flammable liquid is defined as a liquid that can catch fire. The flash point of a flammable liquid is the lowest temperature at which it can form an ignitable mixture with air and produce a flame when a source of ignition is present. Flammable liquids are chemicals that have a flash point below 100°F (38.7°C) and a vapor pressure that does not exceed 40 psig at 100°F. Less-flammable liquids (with a flash point between 100°F and 200°F) are defined as combustible liquids.

A1.1.2 Hazardous Chemical(s) or Class of Hazardous Chemical(s)

Flammable liquids are commonly divided into three classes.

Class	Flash Point	Boiling Point	Example
IA	Below 73°F	Below 100°F	Ethyl ether
IB	Below 73°F	At or above 100°F	Acetone, Benzene, Toluene
IC	At or above 73°F and below 100°F		Methanol, Isopropanol, Xylene

Combustible liquids are divided into three classes.

Class	Flash Point	Example
II	100–139°F	Acetic acid, cyclohexane, and mineral spirits
IIIA	140–199°F	Cyclohexanol, formic acid, and nitrobenzene
IIIB	200°F or above	Formalin and vegetable oil

A1.1.3 Personal Protective Equipment

A1.1.3.1 Eye Protection

ANSI compliant safety glasses with side shields should be worn. Goggles should be worn when working with larger quantities. If chemical has a skin hazard or is a caustic liquid, a face shield should be worn when splashing onto the face is a possibility.

A1.1.3.2   Skin and Body Protection

Wear a flame-resistant lab coat. Laboratory coats must be appropriately sized for the individual and be buttoned to their full length. Laboratory coat sleeves must be of a sufficient length to prevent skin exposure while wearing gloves. A chemical-resistant apron should be used when transferring or using large quantities and splashing is a possibility. Wear long pants and closed-toe shoes. Nonsynthetic clothing should be worn.

A1.1.3.3   Hand Protection

At a minimum, wear a nitrile chemical-resistant glove. Consult with your preferred glove manufacturer to ensure that the gloves you plan on using are compatible with the chemical and usage. Additional personal protective equipment (PPE) may be required if procedures or processes present additional risk.

A1.1.4   Engineering/Ventilation Controls

All chemicals should be transferred and used in a laboratory chemical fume hood with the sash at the certified position or lower. The hood flow alarm should be checked to be operating correctly prior to using the hood.

- Safety shielding: Shielding is required any time if there is a risk of explosion, splash hazard, or a highly exothermic reaction. All manipulations of flammable liquids that pose this risk should occur in a fume hood with the sash in the lowest feasible position. Portable shields may also be used.
- Special ventilation: Manipulation of flammable liquids outside of a fume hood may require special ventilation controls to minimize exposure to the material. Fume hoods provide the best protection against exposure to flammable liquids in the laboratory and are the preferred ventilation control device. Always attempt to handle quantities of flammable liquids >500 mL in a fume hood.
- Vacuum protection: Evacuated glassware can implode and eject flying glass and chemicals. Vacuum work involving flammable liquids must be conducted in a fume hood, glove box, or isolated in an acceptable manner. Mechanical vacuum pumps must be protected using cold traps and, where appropriate, filtered to prevent particulate release. The exhaust for the pumps must be vented into a fume hood. Vacuum pumps should be rated for use with flammable liquids.

A1.1.5   Special Handling Procedures and Storage Requirements

Use in an area that is properly equipped with a certified eye wash/safety shower and is available within 10 s of travel. Wash thoroughly after handling. Do not ingest or inhale nor get in eyes, skin, or clothing. Remove contaminated clothing and wash before reuse. Store in a tightly closed, labeled container and in a cool, dry well-ventilated area. Segregate from incompatible materials. Secondary containers must be labeled clearly. Follow any substance-specific storage guidance provided in Safety Data Sheet documentation. Use small quantities whenever possible. Monitor your inventory closely to assure that you have tight control over your material.

A1.1.5.1 Flammable Liquid Storage Cabinets

One or more flammable liquid storage cabinets (FLSCs) are required for laboratories that store, use, or handle more than 5 gallons of flammable or combustible liquids. Containers 1 gallon and larger of flammable liquids must be stored in a flammable-liquids storage cabinet. The storage of flammable and combustible liquids in a laboratory, shop, or building area must be kept to the minimum needed for research and/or operations. FLSCs are not intended for the storage of highly toxic materials, acids, bases, compressed gases, or pyrophoric chemicals. In many laboratories, flammable liquids storage is provided under the chemical fume hood. These cabinets are clearly marked "flammable storage." Flammable liquids storage cabinets are equipped with a grounding system that should be connected to a building ground. If you are pouring from a container in the storage cabinet and if the container being poured into is conductive then a bonding strap must be attached between them. Flammable liquids storage cabinets are constructed to limit the internal temperature when exposed to fire. When additional storage is needed, NFPA 30-4.3.3 approved FLSC may be used.

All containers of flammable liquids must be stored in an FLSC when not in use. The following requirements apply: cabinets shall be no larger than 45 gallon capacity; cabinets should be located near fume hood alcoves; cabinets shall be marked "Flammable-Keep Fire Away"; cabinets should be kept in good condition. Doors that do not close and latch must be repaired or the cabinet must be replaced.

A1.1.5.2 Transferring/Dispensing

A1.1.5.2.1 Static Electricity Hazards in the Laboratory

The flow of flammable and combustible liquids can cause the buildup of static electricity. When enough of a charge is built up a spark can result and potentially cause a fire or explosion. The likelihood of this happening is dependent upon how well the liquid conducts electricity, the flash point, and the capacity to generate static electricity. Static electricity can be generated when liquid is transferred from one metal container to another. Liquids have the ability to generate static electricity when they move in contact with other materials during pouring, pumping, or agitating. The buildup of this static electricity can cause a spark to form where the solvent exits the container. This spark could result in a fire or explosion. To avoid the buildup of static electricity that may cause a spark, it is important to bond and ground metal or special conductive plastic containers. Bonding eliminates the electrical potential between two containers, therefore reducing the likelihood of sparks. A bonding wire is connected to two conductive objects as seen in the drums pictured to the left. Grounding eliminates the difference in static potential charge between the conductive object and ground. Grounding is accomplished by connecting the conductive object directly to the earth, usually using cold water copper pipes, building steel or grounding bus/bar. Bonding and grounding require good electrical connections. Remove any dirt, paint, or rust ensuring metal-to-metal contact.

A1.1.5.2.2 Dispensing Flammable Liquids From 5-Gallon Containers

Manual dispensing pumps for 5-gallon cans are available. These pumps are specifically designed to dispense liquids into small laboratory-size bottles without spilling.

If you are dispensing into a conductive container, a bonding wire should be attached from the 5-gallon can to the container being filled. The 5-gallon can should be grounded.

A1.1.5.2.3 Dispensing Flammables From Safety Cans

Safety cans have self-closing air-tight lids and a flame arrester that protects the contents from an external ignition source. Bonding and grounding is still required on safety cans since static electricity generation is possible. The nozzle provides a bonding path to a receiving metallic vessel. If either of the containers are nonmetallic (conductive), it is still important to follow these precautions. Safety cans do not offer protection from heat when exposed to fire and should be stored in a flammable liquids storage cabinet when not in use.

A1.1.5.3 Labeling

All flammable liquids must be clearly labeled with the correct chemical name. Handwritten labels are acceptable; chemical formulas and structural formulas are not acceptable. The label on any containers of flammable liquids should say "Flammable" and include any other hazard information, such as "Corrosive" or "Toxic," as applicable.

A1.1.5.4 Heating

Do not store flammable liquids in chemical fume hoods or allow containers of flammable liquids in proximity to heating mantles, oil baths, or hot plates. With the exception of vacuum drying ovens, laboratory ovens rarely have any means of preventing the discharge of material volatilized within them. Thus it should be assumed that flammable liquid residues in items placed within the oven will escape into the laboratory atmosphere, and may also be present in sufficient concentration to form explosive mixtures within the oven itself. Venting the oven to the exhaust system will reduce this hazard. Drying ovens should not be used to dry glassware that has been rinsed with organic solvents until the majority of the solvent has had the opportunity to drain or evaporate at room temperature.

A1.2 Acids

A1.2.1 Hazard Overview

Acids are corrosive to eyes, skin, and mucous membranes. Acid burns are immediately painful due in part to the formation of a protein layer that resists further penetration of the acid. Corrosive effects can occur not only on the skin and eyes, but also in the respiratory tract and, in the case of ingestion, in the gastrointestinal tract as well.

The pH range of acids is 0–6.9 (water = 7.0 = neutral). A pH of approximately 0–3 represents a strong acid. Some inorganic acids fall within this range. Weak acids (pH of 3–7) include diluted acetic acid solutions and boric acid. Weak acids irritate the skin with short contact and can cause burns with prolonged contact.

Heat is released when strong acids are mixed with water. When water is added to acid, an extremely concentrated acid solution is formed initially and the solution may boil very violently, splashing concentrated acid. Additionally, because water is less dense than most acids, it may not mix into the acid easily, but will sit on top of the acid solution. The heat of solution can cause the more volatile aqueous acid solution to boil and escape the container. When acid is added to water, the solution formed is dilute and the small amount of heat released in each addition increment is not significant to vaporize and spatter it.

Concentrated aqueous solutions of inorganic acids are not in themselves flammable. Acids also react with many metals, resulting in the liberation of hydrogen, a highly flammable gas. Some acids are strong oxidizing agents and can react destructively and violently when in contact with other materials. For this reason, it is essential to read the warning labels on the acid in question indicating physical and chemical hazards.

A1.2.2   Hazardous Acids

All acid solutions are considered hazardous.

A1.2.3   Personal Protective Equipment

A1.2.3.1   Eye Protection

At a minimum, approved safety glasses with side shields should be worn. Chemical splash goggles should be worn when working with larger quantities or concentrated aqueous solutions of acids.

A1.2.3.2   Skin and Body Protection

Wear a chemical-resistant lab coat, long pants, and closed-toe shoes. These laboratory coats must be appropriately sized for the individual and be buttoned to their full length. Laboratory coat sleeves must be of a sufficient length to prevent skin exposure while wearing gloves. A chemical-resistant apron should be used when transferring or using large quantities of acids or concentrated aqueous solutions, when acid splashing is a possibility.

A1.2.3.3   Hand Protection

At a minimum, wear a nitrile chemical-resistant glove. Consult with your preferred glove manufacturer to ensure that the gloves you plan on using are compatible with the chemical and usage.

A1.2.4   Engineering/Ventilation Controls

All chemicals should be transferred and used in a laboratory chemical fume hood with the sash at the certified position or lower. The hood flow alarm should be checked to be operating correctly prior to using the hood.

A1.2.5   Special Handling Procedures and Storage Requirements

Acids can be only used in areas properly equipped with a certified eye wash/safety shower that can be reached within 10 s. It is essential that all strong corrosives be stored separately from other laboratory chemicals with which they may react. Ensure secondary containment and segregation of incompatible chemicals. Also, follow any substance-specific storage guidance provided in Safety Data Sheet (SDS) documentation. The corrosive properties of these materials and their ability to produce fires or explosions by combination with combustible materials make the following considerations mandatory in the selection of a storage site.

- Maintain a relatively cool, dry environment free from temperature–humidity extremes.
- Store acids in material that is acid resistant; this facilitates flushing and other cleanup procedures in the event of leaks or spills.
- Store on low shelves or in acid/base storage cabinets.
- When mixing acids and water, always add acid to water. NEVER add water to acid!
- Use bottle carriers for transporting materials whenever possible.
- Segregate oxidizing acids from organic acids, and flammable and combustible liquids.
- Segregate acids from active metals such as sodium, magnesium, zinc, etc.
- Store mineral acids together, separate from oxidizing agents and organic materials.
- Store acetic acid and other organic acids with the combustible organic liquids.

A1.2.6 Spill and Incident Procedures

Assess the extent of danger. Help contaminated or injured persons. Evacuate the spill area. Avoid breathing vapors. If possible, confine the spill to a small area using a spill kit or absorbent material. Keep others from entering contaminated area (e.g., caution tape, barriers, etc.).

Small (<500 mL)—If you have training and do not perceive the risk to be greater than normal laboratory operations, use appropriate PPE and cleanup materials for the chemical spilled. Cover spill with sodium carbonate or bicarbonate (be careful of possible strong reaction). When reaction stops, pick up with damp sponge or paper towels. Place waste in container, label, and arrange for chemical waste pickup.

Large (>500 mL)—Call for help. Notify others in area of spill. Turn off ignition sources in area. Evacuate area and post doors to spill area. Remain on the scene, but at a safe distance, to receive and direct safety personnel when they arrive.

Chemical spill on body or clothes—Remove clothing and rinse body or affected skin with plenty of water or thoroughly in emergency shower for at least 15 min. Seek medical attention.

Chemical splash into eyes—Immediately rinse eyeball and inner surface of eyelid with water from the emergency eyewash station for 15 min by forcibly holding the eye open. Seek medical attention.

A1.2.7 Decontamination

Wearing proper PPE, decontaminate equipment and bench tops using sodium bicarbonate and water. Dispose of all used contaminated disposables as hazardous waste following the Waste Disposal Section.

A1.2.8 Waste Disposal

All acid waste must be disposed of as hazardous waste. General disposal guidelines: affix to all waste containers a hazardous waste tag; store hazardous waste in closed containers, in secondary containment, and in a designated location; do not let waste enter the building drains, as discharge into the environment must be prevented; double-bag dry waste using transparent bags; waste must be under the control of the person/laboratory during its generation and until disposal; dispose of routinely generated chemical waste within 90 days.

A1.2.9 Prior Approval/Review Required

Prior to use of acids, the following shall be completed: documented specific training on the techniques and processes to be used; read and understand the relevant Safety Data Sheet; demonstrate competence to perform work; review this document when there are any changes to procedures, personnel, equipment, or when an incident or near miss occurs.

A1.2.10 Designated Area

Work should be completed in a laboratory fume hood. All hoods in our lab are designated for acids.

A1.2.11 Safety Data Sheets

SDSs can be found online.

A1.2.12 Detailed Protocol

The most common uses of acids are for reaction neutralization and solvent extractions. For these purposes, dilute aqueous acid solutions (up to 1 M) are mostly used. For volumes of these solutions up to 100 mL per use, conventional laboratory precautions should be sufficient. For volumes larger than this, or for the preparation of dilute solutions from more concentrated reagents, greater PPE should be used, including mandatory use of goggles. Laboratory personnel must have specific hands-on training on the proper handling of acid and acid solutions and understanding the hazards. Laboratory personnel using acids or acid solutions must demonstrate competence by being able to (1) list the foreseeable emergency situations, (2) describe the proper response to the emergency situations, and (3) know the control measures to minimize the risks. The research laboratory requires variation in reaction conditions to develop and optimize new chemical transformations. The researcher must seek literature precedence or have relevant personal experience for reaction conditions that have reasonable similarities to new chemistry that is planned with the acids described here. The researcher must also receive approval to proceed with chemical transformations that have little literature or local research group precedence. Approval must also be obtained for scale-up of such transformations greater than a factor of 5.

When working in the lab, a worker must: not work alone; not use quantities of acid solutions >500 mL in any given reaction without prior approval; be cognizant of all SDS and safety information presented in this document; follow all related procedures in the laboratory (PPE, syringe techniques, waste disposal, etc. as appropriately modified by any specific information in the SDS information presented in this document); discuss all issues or concerns regarding acid solutions with the laboratory supervisor prior to use.

If there is an unusual or unexpected occurrence when using acid solutions, the occurrence must be documented and discussed with the laboratory supervisor and others who might be using acid solutions. Unusual or unexpected occurrences might include a fire, explosion, sudden rise or drop in temperature, increased rate of gas evolution, color change, phase change, or separation into layers.

A1.3   Bases

A1.3.1   Hazard Overview

Bases are corrosive and will destroy body tissue. The extent of injury depends on factors such as the type and concentration of the base, the route of exposure, the type of tissue contacted, and the speed used in applying emergency measures. Skin contact with strong bases usually goes unnoticed, since pain does not occur immediately.

The eyes are especially susceptible to bases and must be immediately flushed with water for at least 15 min if exposure occurs. Inhaling airborne dust and mist from bases irritate the nose, throat, and lungs. Pulmonary edema, a severe irritation of the lungs resulting in fluid production that prevents the transfer of oxygen to the bloodstream, can also occur from intense extreme airborne exposures. Secondary toxic effects may occur if the material is absorbed from the lungs into the bloodstream. The extent of these effects depends on the concentration in air and the duration of exposure.

Dilution of bases is exothermic. This is particularly true for potassium hydroxide. Concentrated solutions of inorganic bases are not in themselves flammable. Bases such as sodium hydroxide will liberate hydrogen gas upon reaction with aluminum, magnesium, tin, and zinc metal.

A1.3.2   Hazardous Bases

All base solutions are considered hazardous.

A1.3.3   Personal Protective Equipment

A1.3.3.1   Eye Protection

At a minimum, approved safety glasses with side shields should be worn. Chemical splash goggles should be worn when working with larger quantities or concentrated aqueous solutions of acids.

A1.3.3.2   Skin and Body Protection

Wear a chemical-resistant lab coat, long pants, and closed-toe shoes. These laboratory coats must be appropriately sized for the individual and be buttoned to their full length. Laboratory coat sleeves must be of a sufficient length to prevent skin exposure while wearing gloves. A chemical-resistant apron should be used when transferring or using large quantities of acids or concentrated aqueous solutions, when acid splashing is a possibility.

A1.3.3.3   Hand Protection

At a minimum, wear a nitrile chemical-resistant glove. Consult with your preferred glove manufacturer to ensure that the gloves you plan on using are compatible with the chemical and usage.

A1.3.4   Engineering/Ventilation Controls

All concentrated bases should be transferred and used in a laboratory chemical fume hood with the sash at the certified position or lower. The hood flow alarm should be checked to be operating correctly prior to using the hood.

A1.3.5   Special Handling Procedures and Storage Requirements

Bases can be only used in areas properly equipped with a certified eye wash/safety shower that can be reached within 10 s. Store in a tightly closed, labeled container and in a cool, dry well-ventilated area. Ensure secondary containment and segregate from incompatible materials. Secondary containers must be labeled clearly. Follow any substance-specific storage guidance provided in Safety Data Sheet documentation. Use small quantities whenever possible. Monitor your inventory closely to assure that you have tight control over your material.

The corrosive properties of these materials and their ability to produce fires or explosions by combination with combustible materials make the following considerations mandatory in the selection of a storage site.

- Maintain a relatively cool, dry environment free from temperature–humidity extremes.
- Store bases in material that is base resistant; this facilitates flushing and other cleanup procedures in the event of leaks or spills.
- Store on low shelves or in base storage cabinets.
- Use bottle carriers for transporting materials whenever possible.
- Store solutions of inorganic hydroxides in polyethylene containers.

A1.3.6   Spill and Incident Procedures

Assess the extent of danger. Help contaminated or injured persons. Evacuate the spill area. Avoid breathing vapors. If possible, confine the spill to a small area using a spill kit or absorbent material. Keep others from entering contaminated area (e.g., caution tape, barriers, etc.).

Small (<500 mL)—If you have training and do not perceive the risk to be greater than normal laboratory operations, use appropriate PPE and cleanup materials for the chemical spilled. Double bag spill waste in clear plastic bags, label, and arrange for chemical waste pickup.

Large (>500 mL)—Call for help. Notify others in area of spill. Turn off ignition sources in area. Evacuate area and post doors to spill area. Remain on the scene, but at a safe distance, to receive and direct safety personnel when they arrive.

Chemical spill on body or clothes—Remove clothing and rinse body or affected skin with plenty of water or thoroughly in emergency shower for at least 15 min. Seek medical attention.

Chemical splash into eyes—Immediately rinse eyeball and inner surface of eyelid with water from the emergency eyewash station for 15 min by forcibly holding the eye open. Seek medical attention.

A1.3.7 Decontamination

Wearing proper PPE, collect any crystals with a brush—avoid creating dust. Decontaminate equipment and bench tops. Dispose of all used contaminated disposables as hazardous waste following the Waste Disposal Section.

A1.3.8 Waste Disposal

All waste must be disposed of as hazardous waste. General disposal guidelines: affix to all waste containers a hazardous waste tag; store hazardous waste in closed containers, in secondary containment, and in a designated location; do not let waste enter the building drains, as discharge into the environment must be prevented; double-bag dry waste using transparent bags; waste must be under the control of the person/laboratory during its generation and until disposal; dispose of routinely generated chemical waste within 90 days.

A1.3.9 Prior Approval/Review Required

Prior to use of bases, the following shall be completed: documented specific training on the techniques and processes to be used; read and understand the relevant Safety Data Sheet; demonstrate competence to perform work; review this document when there are any changes to procedures, personnel, equipment, or when an incident or near miss occurs.

A1.3.10 Designated Area

Work should be completed in a laboratory fume hood. All hoods in our lab are designated for bases.

A1.3.11 Safety Data Sheets

SDSs can be found online.

A1.3.12   Detailed Protocol

The most common uses of bases or basic solutions are for reaction neutralization and solvent extractions. For these purposes, dilute base solutions (up to 1 M) are mostly used. For volumes of these solutions up to 100 mL per use, conventional laboratory precautions should be sufficient. For volumes larger than this, or for the preparation of dilute solutions from more concentrated reagents, greater PPE should be used, including mandatory use of goggles. Laboratory personnel must have specific hands-on training on the proper handling of base and base solutions and understanding the hazards. Laboratory personnel using base or base solutions must demonstrate competence by being able to (1) list the foreseeable emergency situations, (2) describe the proper response to the emergency situations, and (3) know the control measures to minimize the risks. The research laboratory requires variation in reaction conditions to develop and optimize new chemical transformations. The researcher must seek literature precedence or have relevant personal experience for reaction conditions that have reasonable similarities to new chemistry that is planned with the acids described here. The researcher must also receive approval to proceed with chemical transformations that have little literature or local research group precedence. Approval must also be obtained for scale-up of such transformations greater than a factor of 5.

When working in the lab, a worker must: not work alone; not use quantities of base solutions >500 mL in any given reaction without prior approval; be cognizant of all SDS and safety information presented in this document; follow all related procedures in the laboratory (PPE, syringe techniques, waste disposal, etc. as appropriately modified by any specific information in the SDS information presented in this document); discuss all issues or concerns regarding base solutions with the laboratory supervisor prior to use.

If there is an unusual or unexpected occurrence when using base solutions, the occurrence must be documented and discussed with the laboratory supervisor and others who might be using base solutions. Unusual or unexpected occurrences might include a fire, explosion, sudden rise or drop in temperature, increased rate of gas evolution, color change, phase change, or separation into layers.

A1.4   Peroxide-Forming Chemicals

A1.4.1   Hazard Overview

Organic peroxides are among the most dangerous substances handled in the chemical laboratory. They are generally low-power explosives that are sensitive to shock, sparks, or other accidental ignition. They are far more shock sensitive than most primary explosives such as TNT. Organic peroxides are organic compounds containing the peroxide functional group (ROOR′). These materials are sensitive to oxygen, heat, friction, impact, light, and strong oxidizing and reducing agents.

Peroxide formation may occur when compounds of particular chemical classes are stored for prolonged periods, concentrated through distillation, evaporation or air exposure, and also as a result of polymerization. Peroxide-forming chemicals are compounds that undergo autoxidation to form organic hydroperoxides and/or peroxides when exposed to oxygen (in air or in chemical processes involving gases enriched in oxygen). Especially dangerous are ether bottles that have evaporated to dryness. A peroxide present as a contaminant in a reagent or solvent can be very hazardous. Autoxidation of organic materials (solvents and other liquids used in quantity are of greatest concern) proceeds by a free-radical chain mechanism. For alkane substrates R—H, the chain can be initiated by ultraviolet light. The unusual stability problems of this class of compounds make them a serious fire and explosion hazard that requires careful management.

A1.4.2   Hazardous Chemicals/Class of Hazardous Chemicals

The following are some classes of organic compounds prone to forming peroxides: ethers, especially cyclic and those containing secondary alkyl groups (never distill an ether before it has been shown to be free of peroxide); aldehydes; compounds containing benzylic hydrogen; compounds containing allylic hydrogens (C=C—CH), including most alkenes; vinyl and vinylidene compounds; compounds containing a tertiary C—H group (e.g., decalin and 2,5-dimethyl hexane).

A1.4.3   Personal Protective Equipment

A1.4.3.1   Eye Protection

At a minimum, approved safety glasses with side shields should be worn. Chemical splash goggles should be worn when working with larger quantities or concentrated aqueous solutions of acids.

A1.4.3.2   Skin and Body Protection

Wear a chemical-resistant lab coat, long pants, and closed-toe shoes. These laboratory coats must be appropriately sized for the individual and be buttoned to their full length. Laboratory coat sleeves must be of a sufficient length to prevent skin exposure while wearing gloves. A chemical-resistant apron should be used when transferring or using large quantities of acids or concentrated aqueous solutions, when acid splashing is a possibility.

A1.4.3.3   Hand Protection

At a minimum, wear a nitrile chemical-resistant glove. Consult with your preferred glove manufacturer to ensure that the gloves you plan on using are compatible with the chemical and usage.

A1.4.4   Engineering/Ventilation Controls

Use one of the following engineering controls. (1) Fume hood: Work inside a certified chemical fume hood at all times. A portable safety shield may also be used to control the risk from explosion. (2) Glove box: When inert or dry atmospheres are required.

Know where your safety equipment is located (i.e., fire extinguisher, eye wash/ safety shower, and first aid kit). Have the appropriate fire extinguisher available.

A1.4.5 Special Handling Procedures and Storage Requirements

Design your experiment to use the least amount of peroxide-forming material possible to achieve the desired result. Never work alone. At least one other person must be present in the same laboratory when any work involving peroxide-forming chemicals is carried out. Eliminate or substitute a less hazardous material when possible. Verify your experimental setup and procedure prior to use. Ensure all equipment is appropriate for the task. Consult with a chemical hygiene officer if work involves large quantities of peroxide-forming material. Establish a designated area: clear the area of unrelated and possibly incompatible hazards; keep container sizes and quantities in the work area as small as possible; transport material in secondary containers and only in small quantities; only use if the area is properly equipped with a certified eye wash/ safety shower reachable within 10 s; avoid incompatible hazards—heat sources, open flames, and oxidizers.

Diethyl ether may be used only in a fume hood. THF-containing HPLC mobile phases must be prepared in the fume hood but may be used outside of the fume hood on HPLC equipment so long as the mobile phase supply container is sealed. Refrigeration of diethyl ether is not recommended. Reduced temperature can impede the peroxide-scavenging ability of added preservatives and actually increase peroxide formation.

Purchase and use the minimum amount of material necessary to perform your research. Label peroxide-forming materials clearly and promptly upon receipt or synthesis. Store all peroxide-forming materials inside of a flammable cabinet. Review your inventory frequently to prevent peroxide-forming chemicals from becoming unsafe. Test materials for peroxide formation according to guidance set forth in the SDS. Do not handle old or expired peroxide-forming materials that are discovered. Inform your laboratory supervisor immediately and dispose of the item as a hazardous waste.

Ether solvents: Ether solvents stored in solvent drying cartridge manifolds can be excluded from the most stringent precautions since these are kept air-free under a positive pressure of inert gas. The dangers associated with ether solvents depend on and can be exacerbated by these factors: exposure to oxygen; exposure to light; temperature; friction; shock; concentration; chemical structure; slow evaporation of volatile ethers over time.

A1.4.6 Spill and Incident Procedures

Assess the extent of danger. Help contaminated or injured persons. Evacuate the spill area. Avoid breathing vapors. If possible, confine the spill to a small area using a spill kit or absorbent material. Keep others from entering contaminated area (e.g., caution tape, barriers, etc.).

Small (<500 mL)—If you have training and do not perceive the risk to be greater than normal laboratory operations, use appropriate PPE and cleanup materials for the chemical spilled. Double bag spill waste in clear plastic bags, label, and arrange for chemical waste pickup.

Large (>500 mL)—Call for help. Notify others in area of spill. Turn off ignition sources in area. Evacuate area and post doors to spill area. Remain on the scene, but at a safe distance, to receive and direct safety personnel when they arrive.

Chemical spill on body or clothes—Remove clothing and rinse body or affected skin with plenty of water or thoroughly in emergency shower for at least 15 min. Seek medical attention.

Chemical splash into eyes—Immediately rinse eyeball and inner surface of eyelid with water from the emergency eyewash station for 15 min by forcibly holding the eye open. Seek medical attention.

A1.4.7   Decontamination

If there are incidental drips of diethyl ether or THF on the fume hood work surface, secure ignition sources and lower the sash to allow for evaporation.

A1.4.8   Waste Disposal

All waste must be disposed of as hazardous waste by trained lab workers. General disposal guidelines: affix to all waste containers an online hazardous waste tag; store hazardous waste in closed containers, in secondary containment, and in a designated location; do not let waste enter the building drains, as discharge into the environment must be prevented; double-bag dry waste using transparent bags; waste must be under the control of the person/laboratory during its generation and until disposal; dispose of routinely generated chemical waste within 90 days.

A1.4.9   Prior Approval/Review Required

All work with peroxide-forming chemicals require the following prior to beginning work: documented specific training on the techniques and processes to be used; read and understand the relevant Safety Data Sheet; demonstrate competence to perform work; review this document when there are any changes to procedures, personnel, equipment, or when an incident or near miss occurs. The laboratory supervisor must be notified if there is any suspicion that any material in the laboratory has formed even small quantities of peroxides. It is crucial that no such material be heated above room temperature or the solvent volume reduced (e.g., by rotary evaporation under reduced pressure) without the express permission of the supervisor.

A1.4.10   Designated Area

Work should be completed in a laboratory fume hood. All hoods are designated for peroxide-forming chemicals.

A1.4.11 Safety Data Sheets

SDSs can be found online. Additionally, review R. J. Kelly's paper *Review of Safety Guidelines for Peroxidizable Organic Chemicals* (*Journal of Chemical Health and Safety*; Sept/Oct 1996).

A1.4.12 Detailed Protocol

A surprisingly large number of organic compounds react spontaneously with O_2 in the air to form peroxides. Butadiene (and likely other dienes) and isopropyl ether (and likely other ethers with tertiary α-hydrogens) can form explosive levels of peroxides even without concentration by evaporation/distillation. A large number of compounds can form explosive levels of peroxides upon concentration. They include acetaldehyde, diacetylene, benzyl alcohol, 2-butanol, cumene, cyclohexanol, 2-cyclohexen-1-ol, cyclohexene, decahydronaphthalene, methylcyclopentane, dicyclopentadiene, diethyl ether, dioxanes, glyme, diglyme, 2-hexanol, 4-heptanol, 3-methyl-1-butanol, methyl isobutyl ketone, 4-methyl-2-pentanol, 2-pentanol, 4-penten-1-ol, 1-phenylethanol, 2-phenylethanol, 2-propanol, tetrahydrofuran, and tetrahydronaphthalene.

Peroxide-forming chemicals should be stored in the original manufacturer's container whenever possible. This is very important in the case of diethyl ether because the iron in the steel containers in which it is shipped acts as a peroxide inhibitor. In general, peroxide-forming chemicals should be stored in sealed, air-impermeable containers and should be kept away from light, which can initiate peroxide formation. Dark amber glass with a tight-fitting cap is recommended. Peroxide-forming chemicals can be stored for 3 months after opening the container if they form peroxides without concentration, and for 12 months after opening the containers if they form peroxides with concentration. Materials may be retained beyond this suggested shelf life only if they have been tested for peroxides (see later), show peroxide concentrations <100 ppm, and are retested frequently.

All solvents that are to be distilled should be tested for the presence of peroxides regardless of their age. A safe level for peroxides is considered to be <100 ppm. While several methods are available to test for peroxides in the laboratory, the most convenient is the use of peroxide test strips available from many chemical suppliers. For volatile organic chemicals, the test strip is immersed in the chemical for 1 s. The chemist breathes on the strip for 15–30 s or until the color stabilizes, and the color is compared with a provided colorimetric scale. Any container found to have a peroxide concentration ≥100 ppm should be disposed of.

Researchers should never test containers of unknown age or origin. Older containers are far more likely to have concentrated peroxides or peroxide crystallization in the cap threads and therefore can present a serious hazard when opened for testing.

Laboratory personnel using peroxide-forming chemicals must demonstrate competence to the laboratory supervisor by being able to (1) list the foreseeable emergency situations, (2) describe the proper response to the emergency situations, and (3) know the control measures to minimize the risks.

If there is an unusual or unexpected occurrence when using the peroxide-forming chemicals, the occurrence must be documented and discussed with the laboratory supervisor and others who might be using peroxide-forming chemicals. Unusual or unexpected occurrences might include a fire, explosion, sudden rise or drop in temperature, increased rate of gas evolution, color change, phase change, or separation into layers.

A1.5   Strong Oxidizing Agents

A1.5.1   Hazard Overview

Oxidizing materials are liquids or solids that readily give off oxygen or other oxidizing substances (such as bromine, chlorine, or fluorine). They also include materials that react chemically to oxidize combustible (burnable) materials; this means that oxygen combines chemically with the other material in a way that increases the chance of a fire or explosion. This reaction may be spontaneous at either room temperature or may occur under slight heating. Oxidizing liquids and solids can be severe fire and explosion hazards.

The (US) National Fire Protection Association (NFPA) Code 430 (1995) "Code for the Storage of Liquid and Solid Oxidizers" has classified oxidizing materials according to their ability to cause spontaneous combustion and how much they can increase the burning rate.

Class 1 oxidizers: slightly increase the burning rate of combustible materials; do not cause spontaneous ignition when they come in contact with them.

Class 2 oxidizers: increase the burning rate of combustible materials moderately with which they come in contact; may cause spontaneous ignition when in contact with a combustible material.

Class 3 oxidizers: severely increase the burning rate of combustible materials with which they come in contact; will cause sustained and vigorous decomposition if contaminated with a combustible material or if exposed to sufficient heat.

Class 4 oxidizers: can explode when in contact with certain contaminants; can explode if exposed to slight heat, shock, or friction; will increase the burning rate of combustibles; can cause combustibles to ignite spontaneously.

A1.5.2   Hazardous Chemical(s) or Class of Hazardous Chemical(s)

The NFPA Code 430 (1995) "Code for the Storage of Liquid and Solid Oxidizers" provides many examples of typical oxidizing materials listed according to the NFPA classification system.

Examples of NFPA class 1 oxidizers include aluminum nitrate; ammonium persulfate; barium peroxide; hydrogen peroxide solutions (8–27.5% by weight); magnesium nitrate; nitric acid (40% concentration or less); perchloric acid solutions (less than 50% by weight); potassium dichromate; potassium nitrate; silver nitrate; sodium dichloroisocyanurate dihydrate; sodium dichromate; sodium persulfate; sodium nitrate; sodium nitrite; sodium perborate (and its monohydrate); strontium nitrate; strontium peroxide; trichloroisocyanuric acid; zinc peroxide.

Examples of NFPA class 2 oxidizers include calcium chlorate; calcium hypochlorite (50% or less by weight); chromic acid (chromium trioxide); 1,3-dichloro-5,5-dimethylhydantoin; hydrogen peroxide (27.5–52% by weight); magnesium perchlorate; nitric acid (concentration greater than 40% but less than 86%); potassium permanganate; sodium permanganate; sodium chlorite (40% or less by weight); sodium perchlorate (and its monohydrate); sodium peroxide.

Examples of NFPA class 3 oxidizers include ammonium dichromate; hydrogen peroxide (52–91% by weight); nitric acid, fuming (concentration greater than 86%); perchloric acid solutions (60–72% by weight); potassium bromate; potassium chlorate; potassium dichloroisocyanurate; sodium chlorate; sodium chlorite (greater than 40% by weight); sodium dichloroisocyanurate.

Examples of NFPA class 4 oxidizers include ammonium perchlorate (particle size greater than 15 µm); ammonium permanganate; hydrogen peroxide (greater than 91% by weight); perchloric acid solutions (greater than 72.5% by weight); tetranitromethane.

A1.5.3 Personal Protective Equipment

A1.5.3.1 Eye Protection

At a minimum, approved safety glasses with side shields should be worn. Chemical splash goggles should be worn when working with larger quantities or concentrated aqueous solutions of acids.

A1.5.3.2 Skin and Body Protection

Wear a chemical-resistant lab coat, long pants, and closed-toe shoes. These laboratory coats must be appropriately sized for the individual and be buttoned to their full length. Laboratory coat sleeves must be of a sufficient length to prevent skin exposure while wearing gloves. A chemical-resistant apron should be used when transferring or using large quantities of acids or concentrated aqueous solutions, when acid splashing is a possibility.

A1.5.3.3 Hand Protection

At a minimum, wear a nitrile chemical-resistant glove. Consult with your preferred glove manufacturer to ensure that the gloves you plan on using are compatible with the chemical and usage.

A1.5.4 Engineering/Ventilation Controls

All chemicals should be transferred and used in a laboratory chemical fume hood with the sash at the certified position or lower. The hood flow alarm should be checked to be operating correctly prior to using the hood.

The following is general for all strong oxidizers: always use strong oxidizers in a certified chemical fume hood to minimize the potential for the spread of a fire if one should occur; have a fire extinguisher at hand (not just hanging on the wall in a distant part of the laboratory) when working with class 2–4 materials; class 4 oxidizers may not be used without explicit written permission from the laboratory supervisor; if class 4 oxidizers are used, operations MUST be carried out with the addition of a blast shield in a fume hood. No part of the body (for example, hands) should ever be directly exposed to these materials when they are mixed with other chemicals.

Perchloric acid is notorious for causing unanticipated explosions. Perchloric acid can form explosive salts almost anywhere, including in the exhaust ducts of fume hoods and even laboratory benches where other materials have been spilled in the past. Spills should be immediately and thoroughly cleaned up. Many perchlorate salts are shock sensitive and can lay dormant for very long periods. For these reasons, it is imperative that perchloric acid only be used in a specifically designated fume hood. This fume hood shall be prominently marked for use with perchloric acid.

A1.5.5 Special Handling Procedures and Storage Requirements

It is essential that all strong oxidizers be stored separately from all chemicals with which they may react. Store in a tightly closed, labeled container and in a cool, dry well-ventilated area. Ensure secondary containment and labeled secondary containers clearly. Follow any substance-specific storage guidance provided in SDS documentation. Use small quantities whenever possible. Monitor your inventory closely to assure that you have tight control over your material.

A1.5.6 Spill and Incident Procedures

Spills should be immediately and thoroughly cleaned up. Proceed only if you will not injure yourself or others, and it is not an emergency and not likely to become an emergency. Work areas should be demarked with an indicator that strong oxidizing materials are in use or exposed. Keep a chemical spill kit easily accessible. In the case of strong oxidizing agents, consider the likelihood of fire. Do not use organic-based absorbents such as sawdust, especially with perchloric acid.

Small (<100 g)—If you have training, use appropriate PPE and cleanup materials for chemical spilled. Double bag spill waste in clear plastic bags, label, and arrange for chemical waste pickup.

Large (>100 g)—Call for assistance. Notify others in area of spill. Turn off ignition sources in area. Evacuate area and post doors to spill area. Remain on the scene, but at a safe distance, to receive and direct safety personnel when they arrive.

Chemical spill on body or clothes—Remove clothing and rinse body or affected skin with plenty of water or thoroughly in emergency shower for at least 15 min. Seek medical attention.

Chemical splash into eyes—Immediately rinse eyeball and inner surface of eyelid with water from the emergency eyewash station for 15 min by forcibly holding the eye open. Seek medical attention.

A1.5.7 Decontamination

Wear proper PPE. Carefully inspect work areas to make sure no materials remain. Be sure all ignition sources are secured before beginning cleaning up with flammable liquids. Decontaminate equipment and bench tops using wipers moistened with dry, nonpolar solvent (hexane). Dispose of all used contaminated disposables as hazardous waste following the Waste Disposal Section.

A1.5.8 Waste Disposal

All waste must be disposed of as hazardous waste by trained lab workers. General disposal guidelines: affix to all waste containers an online hazardous waste tag; store hazardous waste in closed containers, in secondary containment, and in a designated location; do not let waste enter the building drains, as discharge into the environment must be prevented; double-bag dry waste using transparent bags; waste must be under the control of the person/laboratory during its generation and until disposal; dispose of routinely generated chemical waste within 90 days.

A1.5.9 Prior Approval/Review Required

All work with strong oxidizers require the following prior to beginning work: documented specific training on the techniques and processes to be used; read and understand the relevant Safety Data Sheet; read Prudent Practices in the Laboratory: Handling and Management of Chemical Hazards (Section 3.D.2.3–3.D.3.3 Incompatible Chemicals—Other Oxidizers); demonstrate competence to perform work; review this document when there are any changes to procedures, personnel, equipment, or when an incident or near miss occurs.

A1.5.10 Designated Area

Work should be completed in a laboratory fume hood. All hoods are designated for strong oxidizers.

A1.5.11 Safety Data Sheets

SDSs can be found online.

A1.5.12 Detailed Protocol

The research laboratory requires variation in reaction conditions to develop and optimize new chemical transformations. The researcher must seek literature precedent for reaction conditions that have reasonable similarities to new chemistry that is planned with a strong oxidizing agents described in this SOP. The researcher must also consult the laboratory supervisor or a designated, experienced research coworker for approval to proceed with chemical transformations that have little literature or local research group precedent. Laboratory supervisor approval must also be obtained for fivefold scale-up of new chemical transformations.

A1.6 Strong Reducing Agents

A1.6.1 Hazard Overview

Reductants in chemistry are very diverse. They must be protected, of course, from any inadvertent reaction with oxidizing agents. Strong reductants include the electropositive elemental metals, such as lithium, sodium, magnesium, iron, zinc, and aluminum, even more so when finely divided. These metals donate electrons readily. Hydride transfer reagents, such as $NaBH_4$ and DIBAL-H, are widely used in organic chemistry, primarily in the reduction of carbonyl compounds to alcohols. Metal alloys may also be strong reductants; most are extremely water reactive and in several cases may be considered pyrophoric. Another example of a reducing agent is hydrogen gas (H_2) with a palladium, platinum, or nickel catalyst.

Finally, materials of known reducing potential, such as hydrazine, should also be considered strong reductants. Many such agents may be also characterized as water reactives, pyrophorics, or by their other major hazardous characteristics.

A1.6.2 Hazardous Chemical(s) or Class of Hazardous Chemical(s)

A1.6.2.1 Common Strong Reducing Agents by Group

Electropositive elemental metals: lithium, sodium, potassium, magnesium, iron, zinc, and aluminum; metal alloys: sodium amalgam, zinc–copper couple, aluminum amalgam; hydride transfer reagents: $NaBH_4$, $LiAlH_4$ and diisobutylaluminum hydride (DIBAL-H), diborane; compounds containing the Sn^{+2} ion, such as tin(II) chloride; sulfite compounds; hydrazine.

A1.6.3 Personal Protective Equipment

A1.6.3.1 Eye Protection

At a minimum, approved safety glasses with side shields should be worn. Chemical splash goggles should be worn when working with larger quantities or concentrated aqueous solutions of acids.

A1.6.3.2 Skin and Body Protection

Wear a flame-resistant lab coat, long pants, and closed-toe shoes. These laboratory coats must be appropriately sized for the individual and be buttoned to their full length. Laboratory coat sleeves must be of a sufficient length to prevent skin exposure while wearing gloves. A chemical-resistant apron should be used when transferring or using large quantities of acids or concentrated aqueous solutions, when acid splashing is a possibility.

A1.6.3.3 Hand Protection

At a minimum, wear a nitrile chemical-resistant glove. Consult with your preferred glove manufacturer to ensure that the gloves you plan on using are compatible with the chemical and usage.

A1.6.4 Engineering/Ventilation Controls

All chemicals should be transferred and used in a laboratory chemical fume hood with the sash at the certified position or lower to minimize the potential for the spread of a fire if one should occur. The hood flow alarm should be checked to be operating correctly prior to using the hood. Have an appropriate fire extinguisher at hand (not just hanging on the wall in a distant part of the laboratory) when working with this class of materials. Where available, an inert atmosphere glove box can drastically reduce the chance of inadvertent fire or exposure. If a glove box is not available, then hazardous operations (such as scales larger than 0.1 mol) must be carried out with the addition of a blast shield in a fume hood. No part of the body (for example, hands) should be directly exposed to these materials as they are being mixed with other chemicals.

A1.6.5 Special Handling Procedures and Storage Requirements

Finely divided solids must be transferred under an inert atmosphere. Liquids may be safely transferred by employing syringe techniques and equipment. Clear the hood or work area of strong oxidants, incompatibles, or other extraneous equipment and materials.

It is essential that all strong reducing agents be stored separately from all chemicals with which they may react, in particular, from any oxidizing agents. Ensure secondary containment and segregation of incompatible chemicals.

A1.6.6 Spill and Incident Procedures

Spills should be immediately and thoroughly cleaned up. Proceed only if you will not injure yourself or others, and it is not an emergency and not likely to become an emergency. Work areas should be demarked with an indicator that strong oxidizing materials are in use or exposed. Keep a chemical spill kit easily accessible. In the case of strong reducing agents, consider the likelihood of fire.

Small (<100 g)—If you have training, use appropriate PPE and cleanup materials for chemical spilled. Double bag spill waste in clear plastic bags, label, and arrange for chemical waste pickup.

Large (>100 g)—Call for assistance. Notify others in area of spill. Turn off ignition sources in area. Evacuate area and post doors to spill area. Remain on the scene, but at a safe distance, to receive and direct safety personnel when they arrive.

Chemical spill on body or clothes—Remove clothing and rinse body or affected skin with plenty of water or thoroughly in emergency shower for at least 15 min. Seek medical attention.

Chemical splash into eyes—Immediately rinse eyeball and inner surface of eyelid with water from the emergency eyewash station for 15 min by forcibly holding the eye open. Seek medical attention.

A1.6.7 Decontamination

Wear proper PPE. Carefully inspect work areas to make sure no materials remain. Be sure all ignition sources are secured before beginning cleaning up with flammable liquids. Decontaminate equipment and bench tops using wipers moistened with dry, nonpolar solvent (hexane). Be sure all ignition sources are secured before beginning cleaning up with flammable liquids. Dispose of all used contaminated disposables as hazardous waste following the Waste Disposal Section.

A1.6.8 Waste Disposal

All waste must be disposed of as hazardous waste by trained lab workers. General disposal guidelines: affix to all waste containers an online hazardous waste tag; store hazardous waste in closed containers, in secondary containment, and in a designated location; do not let waste enter the building drains, as discharge into the environment must be prevented; double-bag dry waste using transparent bags; waste must be under the control of the person/laboratory during its generation and until disposal; dispose of routinely generated chemical waste within 90 days.

A1.6.9 Prior Approval/Review Required

All work with strong reducing agents require the following prior to beginning work: documented specific training on the techniques and processes to be used; read and understand the relevant Safety Data Sheet; demonstrate competence to perform work; review this document when there are any changes to procedures, personnel, equipment, or when an incident or near miss occurs.

A1.6.10 Designated Area

Work should be completed in a laboratory fume hood. All hoods are designated for strong reducing agents.

A1.6.11 Safety Data Sheets

SDSs can be found online.

A1.6.12 Detailed Protocol

Some strong reducing agents are rendered less air-sensitive by being dispersed in mineral oil, such as Li. It is a good idea to remove it before the reaction. This is done as follows: the necessary amount of dispersion (figure by the weight %, usually valid to within 5%) is placed in a dry flask under nitrogen containing a stir bar. The dispersion is covered with dry pentane or petroleum ether. The suspension is stirred for a minute and the stirrer is stopped. The reagent is allowed to settle, and the supernatant is removed by syringe or pipette. This procedure is repeated.

Destruction of excess strong reducing agent is one of the most hazardous aspects of their use. For most alkali metals they should be carefully added to an alcohol to make an alkoxide (Caution: hydrogen evolution), the alkoxide solution should then be added to water to make the hydroxide, and then the hydroxide should be destroyed in accord with the procedure for strong bases. K is the most treacherous of the alkali metals, and fires are most common with K. To quench it, add K slowly to anhydrous *tert*-butanol. If the reaction becomes viscous and stirring slows, add more alcohol. When all K is added, stir until all reaction ceases and examine the solution carefully and extensively for unreacted metal. If none is found, dilute the alcohol solution with water, neutralize, and discard appropriately.

The research laboratory requires variation in reaction conditions to develop and optimize new chemical transformations. The researcher must seek literature precedent for reaction conditions that have reasonable similarities to new chemistry that is planned with strong reducing agents described in this procedure. The researcher must also consult the laboratory supervisor or a designated, experienced research coworker for approval to proceed with chemical transformations that have little literature or local research group precedent. Supervisor approval must also be obtained for significant scale-up of new chemistry transformations.

A1.7   Water Reactive Chemicals

A1.7.1   Hazard Overview

Water reactive materials can react violently with water or atmospheric moisture to produce gas and heat. Typical gases produced are H_2, CH_4, H_2S, NH_3, PH_3, HCN, HF, HCl, HF, HI, SO_2, and SO_3. The risks associated with a specific chemical depend on its reactivity and the nature of the gaseous product (flammable, toxic, or both). The mutual production of flammable gas and heat can lead to spontaneous ignition or explosion. Prior to working with any water reactive chemicals, you must identify which gas may be formed in case of exposure to water and learn the risks associated with this gas. The reaction rate of solid material (and therefore heat and gas generation) depends on the material's surface area. Therefore, smaller particle size increases the hazards associated with these materials.

A1.7.2   Hazardous Chemical(s) or Class of Hazardous Chemical(s)

A1.7.2.1   Common Groups of Water Reactive Materials

Grignard reagents: RMgX; alkali metals: Li, Na, K; alkali metal amides: $NaNH_2$; alkali metal hydrides: lithium aluminum hydride; metal alkyls: lithium and aluminum alkyls; chlorosilanes; halides of nonmetals: BCl_3, BF_3, PCl_3, PCl_5, $SiCl_4$, S_2Cl_2; inorganic acid halides: $POCl_3$, $SOCl_2$, SO_2Cl_2; anhydrous metal halides: $AlCl_3$, $AlBr_3$, $TiCl_4$, $ZrCl_4$, $SnCl_4$; organic acid halides and anhydrides of low molecular weight.

Water reactive material may also present additional hazards such as corrosivity or toxicity.

A1.7.3 Personal Protective Equipment

A1.7.3.1 Eye Protection

At a minimum, approved safety glasses with side shields should be worn. Chemical splash goggles should be worn when working with larger quantities or concentrated aqueous solutions of acids.

A1.7.3.2 Skin and Body Protection

Wear a chemical-resistant lab coat, long pants, and closed-toe shoes. These laboratory coats must be appropriately sized for the individual and be buttoned to their full length. Laboratory coat sleeves must be of a sufficient length to prevent skin exposure while wearing gloves. A chemical-resistant apron should be used when transferring or using large quantities of acids or concentrated aqueous solutions, when acid splashing is a possibility.

A1.7.3.3 Hand Protection

At a minimum, wear a nitrile chemical-resistant glove. Consult with your preferred glove manufacturer to ensure that the gloves you plan on using are compatible with the chemical and usage.

A1.7.4 Engineering/Ventilation Controls

All chemicals should be transferred and used in a laboratory chemical fume hood with the sash at the certified position or lower. The hood flow alarm should be checked to be operating correctly prior to using the hood. The following is for all water reactive chemicals: always use water reactive chemicals in a certified chemical fume hood to minimize the potential for the spread of a fire if one should occur; work away from all water sources or potential water splashes; use fresh dry solvents.

A1.7.5 Special Handling Procedures and Storage Requirements

Design a quenching scheme for residual materials prior to using water reactive materials. Never use water to quench the material itself or a reaction where a water-reactive reagent is used. Begin quenching with a low reactivity quenching agent and slowly add more reactive quenching agents. For example, first quench residual sodium metal with isopropanol and then add ethanol to the mixture. Never use water to extinguish fires caused by water reactive materials.

Store in a tightly closed, labeled container and in a cool, dry well-ventilated area. Ensure secondary containment and labeled secondary containers clearly. Follow any substance-specific storage guidance provided in SDS documentation. Use small quantities whenever possible. Monitor your inventory closely to assure that you have tight control over your material.

A1.7.6   Spill and Incident Procedures

Spills should be immediately and thoroughly cleaned up. Proceed only if you will not injure yourself or others, and it is not an emergency and not likely to become an emergency. Keep a chemical spill kit easily accessible. In the case of water reactive chemicals, consider the likelihood of fire.

Small (<100 g)—If you have training, use appropriate PPE and cleanup materials for chemical spilled. Double bag spill waste in clear plastic bags, label, and arrange for chemical waste pickup.

Large (>100 g)—Call for assistance. Notify others in area of spill. Turn off ignition sources in area. Evacuate area and post doors to spill area. Remain on the scene, but at a safe distance, to receive and direct safety personnel when they arrive.

Chemical spill on body or clothes—Remove clothing and rinse body or affected skin with plenty of water or thoroughly in emergency shower for at least 15 min. Seek medical attention.

Chemical splash into eyes—Immediately rinse eyeball and inner surface of eyelid with water from the emergency eyewash station for 15 min by forcibly holding the eye open. Seek medical attention.

A1.7.7   Decontamination

Water reactive by-products and residual feedstock that remains water-reactive must be quenched before presenting for hazardous waste pickup. Wear proper PPE. Carefully inspect work areas to make sure no materials remain. Be sure all ignition sources are secured before beginning cleaning up with flammable liquids. Decontaminate equipment and bench tops using wipers moistened with dry, nonpolar solvent (hexane). Dispose of all used contaminated disposables as hazardous waste following the Waste Disposal Section. Do not flush with water. Cover with dry sand or other noncombustible material. Dispose as hazardous waste following the guidelines described later.

A1.7.8   Waste Disposal

All waste must be disposed of as hazardous waste by trained lab workers. General disposal guidelines: affix to all waste containers an online hazardous waste tag; store hazardous waste in closed containers, in secondary containment, and in a designated location; do not let waste enter the building drains, as discharge into the environment must be prevented; double-bag dry waste using transparent bags; waste must be under the control of the person/laboratory during its generation and until disposal; dispose of routinely generated chemical waste within 90 days.

A1.7.9   Prior Approval/Review Required

All work with water reactive chemicals require the following prior to beginning work: documented specific training on the techniques and processes to be used; read and understand the relevant Safety Data Sheet; demonstrate competence to perform work; review this document when there are any changes to procedures, personnel, equipment, or when an incident or near miss occurs.

A1.7.10   Designated Area

Work should be completed in a laboratory fume hood. All hoods are designated for strong oxidizers.

A1.7.11   Safety Data Sheets

SDSs can be found online.

A1.7.12   Detailed Protocol

The research laboratory requires variation in reaction conditions to develop and optimize new chemical transformations. The researcher must seek literature precedent for reaction conditions that have reasonable similarities to new chemistry that is planned with a strong oxidizing agents described in this SOP. The researcher must also consult the laboratory supervisor or a designated, experienced research coworker for approval to proceed with chemical transformations that have little literature or local research group precedent. Laboratory supervisor approval must also be obtained for fivefold scale-up of new chemical transformations.

A1.8   Pyrophoric Chemicals

A1.8.1   Hazard Overview

This hazard procedure concerns storage, transfer, and use of organolithium reagents. In general, these materials are pyrophoric; they ignite spontaneously when exposed to air. This is the primary hazard and reagents must be handled so as to rigorously exclude air/moisture. They all are dissolved in a flammable solvent. Their other hazards include water reactivity and peroxide formation. They can also be corrosive via the lithium hydroxide that is formed when they come in contact with moisture.

A1.8.2   Hazardous Chemical(s) or Class of Hazardous Chemical(s)

Pyrophoric chemicals include (but are not necessarily limited to) the following:

Methyllithium lithium bromide complex solution
Methyllithium solution purum, ~5% in diethyl ether (~1.6 M)
Methyllithium solution purum, ~1 M in cumene/THF

Methyllithium solution 3.0 M in diethoxymethane
Methyllithium solution 1.6 M in diethyl ether.
Ethyllithium solution 0.5 M in benzene/cyclohexane (9:1)
Isopropyllithium solution 0.7 M in pentane
Butyllithium solution 2.0 M in cyclohexane
Butyllithium solution purum, ~2.7 M in heptane
Butyllithium solution 10.0 M in hexanes
Butyllithium solution 2.5 M in hexanes
Butyllithium solution 1.6 M in hexanes
Butyllithium solution 2.0 M in pentane
Butyllithium solution ~1.6 M in hexanes
Butyllithium solution technical, ~2.5 M in toluene
Isobutyllithium solution technical, ~16% in heptane (~1.7 M)
sec-Butyllithium solution 1.4 M in cyclohexane
tert-Butyllithium solution purum, 1.6–3.2 M in heptane
tert-Butyllithium solution 1.7 M in pentane
(Trimethylsilyl)methyllithium solution 1.0 M in pentane
(Trimethylsilyl)methyllithium solution technical, ~1 M in pentane
Hexyllithium solution 2.3 M in hexane
2-(Ethylhexyl)lithium solution 30–35 wt% in heptane
Lithium acetylide, ethylenediamine complex 90%
Lithium acetylide, ethylenediamine complex 25 wt% slurry in toluene
Lithium (trimethylsilyl)acetylide solution 0.5 M in tetrahydrofuran
Lithium phenylacetylide solution 1.0 M in tetrahydrofuran
Phenyllithium solution 1.8 M in di-n-butyl ether
2-Thienyllithium solution 1.0 M in tetrahydrofuran
Lithium tetramethylcyclopentadienide
Lithium pentamethylcyclopentadienide

A1.8.3   Personal Protective Equipment

A1.8.3.1   Eye Protection

At a minimum, approved safety glasses with side shields should be worn. Chemical splash goggles should be worn when working with larger quantities.

A1.8.3.2   Skin and Body Protection

Wear a flame-resistant lab coat, long pants, and closed-toe shoes. These laboratory coats must be appropriately sized for the individual and be buttoned to their full length. Laboratory coat sleeves must be of a sufficient length to prevent skin exposure while wearing gloves. A chemical-resistant apron should be used when transferring or using large quantities of acids or concentrated aqueous solutions, when acid splashing is a possibility.

A1.8.3.3   Hand Protection

Flame-resistant glove systems must be worn when handling pyrophoric chemicals. These typically include a flame-resistant inner liner under chemically resistant outer gloves. The inner liner can be Kevlar, and the outer gloves neoprene.

A1.8.4   Engineering/Ventilation Controls

All chemicals should be transferred and used in a laboratory chemical fume hood with the sash at the certified position or lower to minimize the potential for the spread of a fire if one should occur. The hood flow alarm should be checked to be operating correctly prior to using the hood. Have an appropriate fire extinguisher at hand (not just hanging on the wall in a distant part of the laboratory) when working with this class of materials. Where available, an inert atmosphere glove box can drastically reduce the chance of inadvertent fire or exposure. If a glove box is not available, then hazardous operations (such as scales larger than 0.1 mol) must be carried out with the addition of a blast shield in a fume hood. No part of the body (for example, hands) should be directly exposed to these materials as they are being mixed with other chemicals.

A1.8.5   Special Handling Procedures and Storage Requirements

Pyrophoric chemicals should be stored under an atmosphere of inert gas or under kerosene as appropriate. Avoid areas with heat/flames, oxidizers, and water sources. Containers carrying pyrophoric materials must be clearly labeled with the correct chemical name and hazard warning. For storage prepare a storage vessel with a septum filled with an inert gas. Select a septum that fits snugly into the neck of the vessel. Dry any new empty containers thoroughly. Insert septum into neck in a way that prevents atmosphere from entering the clean dry (or reagent filled) flask. Insert a needle to vent the flask and quickly inject inert gas through a second needle to maintain a blanket of dry inert gas above the reactive reagent. Once the vessel is fully purged with inert gas, remove the vent needle then the gas line. For long-term storage, the septum should be secured with a copper wire. For extra protection a second same-sized septa (sans holes) can be placed over the first. Use parafilm around the outer septum and remove the parafilm and outer septum before accessing the reagent through the primary septum.

A1.8.6   Spill and Incident Procedures

Powdered lime, dry sand, Celite (diatomaceous earth), or clay-based kitty litter should be used to completely smother and cover any spill that occurs. A container of smotherant should be kept within arm's length when working with a pyrophoric material. If anyone is exposed, or on fire, wash with copious amounts of water. The recommended fire extinguisher is a standard dry powder (ABC) type. Class D extinguishers are recommended for combustible solid metal fires (e.g., sodium, LAH), but not for organolithium reagents.

A1.8.7   Decontamination

No decontamination procedures for pyrophoric materials are required because they do not survive in the open atmosphere of the laboratory. If quenched they produce strong base solutions, which should be treated using the procedures for that hazard class.

A1.8.8   Waste Disposal

A container with any residue of pyrophoric materials should never be left open to the atmosphere. Any unused or unwanted pyrophoric materials must be destroyed by transferring the materials to an appropriate reaction flask for hydrolysis and/or neutralization with adequate cooling. The essentially empty container should be rinsed three times with an inert dry solvent; this rinse solvent must also be neutralized or hydrolyzed. After the container is triple-rinsed, it should be left open in back of a hood or atmosphere at a safe location for at least a week. After the week, the container should then be rinsed three times again.

A1.8.9   Prior Approval/Review Required

All work with pyrophoric chemicals require the following prior to beginning work: documented specific training on the techniques and processes to be used; read and understand the relevant Safety Data Sheet; demonstrate competence to perform work; review this document when there are any changes to procedures, personnel, equipment, or when an incident or near miss occurs.

A1.8.10   Designated Area

A1.8.10.1   Eyewash

Suitable facilities for quick drenching or flushing of the eyes should be within 10 s travel time for immediate emergency use.

A1.8.10.2   Safety Shower

A safety or drench shower should be available within 10 s travel time where pyrophoric chemicals are used.

A1.8.10.3   Fume Hood

Many pyrophoric chemicals release flammable gases and should be handled in a laboratory hood. In addition, some pyrophoric materials are stored under flammable solvents, therefore the use of a fume hood (or glove box) is required to prevent the release of flammable vapors into the laboratory.

A1.8.10.4   Glove (Dry) Box

Glove boxes are an excellent device to control pyrophoric chemicals when inert or dry atmospheres are required.

A1.8.11   Safety Data Sheets

SDSs can be found online.

A1.8.12   Detailed Protocol

A1.8.12.1   Transferring Pyrophoric Reagents With Syringe

In a fume hood or glove box, clamp the reagent bottle to prevent it from moving. Clamp/secure the receiving vessel too. After flushing the syringe with inert gas, depress the plunger and insert the syringe into the Sure/Seal bottle with the tip of the needle below the level of the liquid. Secure the syringe so if the plunger blows out of the syringe body, the plunger and the contents will not impact anyone (aim it toward the back of the containment). Insert a needle from an inert gas source carefully, keeping the tip of the needle above the level of the liquid. Gently open the inert gas flow control valve to slowly add nitrogen gas into the Sure/Seal bottle. This will allow the liquid to slowly fill the syringe (up to 10 mL). Pulling the plunger causes gas bubbles. Let nitrogen pressure push the plunger to reduce bubbles. Excess reagent and entrained bubbles are then forced back into the reagent bottle. The desired volume of reagent in the syringe is quickly transferred to the reaction apparatus by puncturing a rubber septum.

A1.8.12.2   Transferring Pyrophoric Reagents With a Double-Tipped Needle

The double-tipped needle technique is recommended when transferring 50 mL or more. Pressurize the Sure/Seal bottle with nitrogen and then insert the double-tipped needle through the septum into the headspace above the reagent. Nitrogen will pass through the needle. Insert the other end through the septum on a calibrated addition funnel on the reaction apparatus. Push the needle into the liquid in the Sure/Seal reagent bottle and transfer the desired volume. Then withdraw the needle to above the liquid level. Allow nitrogen to flush the needle. Remove the needle first from the reaction apparatus and then from the reagent bottle. For an exact measured transfer, convey from the Sure/Seal bottle to a dry nitrogen flushed graduated cylinder fitted with a double-inlet adapter. Transfer the desired quantity and then remove the needle from the Sure/Seal bottle and insert it through the septum on the reaction apparatus. Apply nitrogen pressure as before and the measured quantity of reagent is added to the reaction flask. To control flow rate, fit a Luer lock syringe valve between two long needles.

A1.9   Explosion Risks

A1.9.1   Hazard Overview

Explosive chemicals can release tremendous amounts of destructive energy rapidly. If not handled properly, these chemicals can pose a serious threat to the health and safety of laboratory personnel, emergency responders, building occupants, chemical waste handlers, and disposal companies.

There are two classes of explosive chemicals. The first is known explosive chemicals that are designed and produced for use as an explosive (e.g., TNT, explosive bolts, bullets, blasting caps, and fireworks). The other class is potentially explosive chemicals

(PECs), which include peroxidizable organic chemicals. Most chemicals that are used in research laboratories are stable and nonexplosive at the time of purchase. Over time, some chemicals can oxidize, become contaminated, dry out, or otherwise destabilize to become PECs (e.g., isopropyl ether, sodium amide, and picric acid).

PECs are chemicals (or combinations thereof) that may cause a sudden release of pressure, gas, and heat when subjected to sudden shock, pressure, or high temperature.

A1.9.2   Hazardous Chemical(s) or Class of Hazardous Chemical(s)

A1.9.2.1   Potentially Explosive Lab Chemicals

These include acetyl peroxide, acetylene, ammonium nitrate, ammonium perchlorate, ammonium picrate, Ba/Pb/Hg azide (heavy metal azides), Li/K/Na azide, organic azides, bromopropyne, butanone peroxide, cumene peroxide, diazodinitrophenol, dinitrophenol, dinitrophenylhydrazine, dinitroresorcinol, dipicryl amine, dipicryl sulfide, dodecanoyl peroxide, ethylene oxide, lauric peroxide, MEK peroxide, mercury fulminate, silver fulminate, nitrocellulose, nitrogen trifluoride, nitrogen triiodide, nitroglycerine, nitroguanidine, nitromethane, nitrourea, picramide, picric acid (trinitrophenol), picryl chloride, picryl sulphonic acid, propargyl bromide (neat), sodium dinitrophenate, succinic peroxide, tetranitroaniline, trinitroaniline, trinitroanisole, trinitrobenzene, trinitrobenzenesulphonic acid, trinitrobenzoic acid, trinitrocresol, trinitronaphthalene, trinitrophenol (picric acid), trinitroresorcinol, trinitrotoluene, urea nitrate.

A1.9.2.2   Potentially Explosive Compound Classes

These include acetylene ($-C{\equiv}C-$), acyl hypohalites (RCO—OX), organic azide (R—N_3), metal azide (M—N_3), azo (—N=N—), diazo (=N=N), diazosulfide (—N=N—S—N=N—), diazonium salts ($RN_2{}^+$), fulminate (—CNO), halogen amine (=N—X), nitrate (—ONO_2), nitro (—NO_2), aromatic or aliphatic nitramine (=N—NO_2) (—NH—NO_2), nitrite (—ONO), nitroso (—NO), ozonides, peracids (—CO—O—O—H), peroxide (—O—O—), hydroperoxide (—O—O—H), metal peroxide (M—O—O—M).

A1.9.2.3   Explosive Salts

These include bromate salts ($BrO_3{}^-$), chlorate salts ($ClO_3{}^-$), chlorite salts ($ClO_2{}^-$), perchlorate salts ($ClO_4{}^-$), picrate salts (2,4,6-trinitrophenoxide), picramate salts (2-amino-4,6-dinitrophenoxide), hypohalite salts (XO^-), iodate salts ($IO_3{}^-$).

A1.9.2.4   Chemicals That May Explode due to Overpressurized Container

These include aluminum chloride, ammonia solution, ammonium hydroxide, ammonium persulfate, anisyl chloride, aqua regia, benzenesulphonyl chloride, bleach, bleaching powder, calcium carbide, calcium hydride, calcium hypochlorite, chloroform, chromic acid, cumene hydroperoxide, cyclohexene, diethyl pyrocarbonate, dimethylamine, formic acid, hydrogen peroxide, lauroyl peroxide, lithium aluminum hydride, lithium hydride, nitric acid, nitrosoguanidine, peracetic acid, phenol, phosphorus trichloride, potassium persulphate, silicon tetrachloride, sodium borohydride, sodium dithionite, sodium hydride, sodium hydrosulphite, sodium hypochlorite, sodium peroxide, sodium persulfate, thionyl chloride, urea peroxide, zinc.

A1.9.3   Personal Protective Equipment

A1.9.3.1   Eye Protection

At a minimum, approved safety glasses with side shields should be worn. Chemical splash goggles should be worn when working with larger quantities or concentrated aqueous solutions of acids.

A1.9.3.2   Skin and Body Protection

Wear a chemical-resistant lab coat, long pants, and closed-toe shoes. These laboratory coats must be appropriately sized for the individual and be buttoned to their full length. Laboratory coat sleeves must be of a sufficient length to prevent skin exposure while wearing gloves. A chemical-resistant apron should be used when transferring or using large quantities of acids or concentrated aqueous solutions, when acid splashing is a possibility.

A1.9.3.3   Hand Protection

At a minimum, wear a nitrile chemical-resistant glove. Consult with your preferred glove manufacturer to ensure that the gloves you plan on using are compatible with the chemical and usage.

A1.9.4   Engineering/Ventilation Controls

All chemicals should be transferred and used in a laboratory chemical fume hood with the sash at the certified position or lower. The hood flow alarm should be checked to be operating correctly prior to using the hood.

- All operations involving PECs should be carried out in a certified fume hood to keep the airborne level below recommended exposure limits.
- Chemical fume hoods that used to contain PECs must have a face velocity of 100 lfm, averaged over the face of the hood.
- Laboratory rooms must be at negative pressure with respect to the corridors and external environment. The laboratory/room door must be kept closed at all times.
- Vacuum lines are to be protected by HEPA (high efficiency particulate air) filters or higher efficiency scrubbers.
- Safety shielding shall be used for any operation having the potential for explosion, including the following situation: when a familiar reaction with a PECs is carried out on a larger than usual scale (i.e., 5–10 times more material). Consult with the laboratory supervisor or designated, experienced individual researcher prior to scale-up.
- Shields must be placed so that all personnel in the area are protected from hazard.

A1.9.5   Special Handling Procedures and Storage Requirements

It is important that chemical users track and dispose of chemicals before they become a problem. Proper inventory management systems can help mitigate risk to personnel and avert higher than normal disposal costs.

- Identify all explosive and PECs in your inventory.
- Record the opening date and the date that the chemical should be discarded on the label of chemicals that may degrade to become potentially explosive.
- Keep explosive chemicals away from all ignition sources such as open flames, hot surfaces, spark sources, and direct sunlight.
- Periodically check containers of chemicals that could become overpressurized, such as highly concentrated formic acid. If there is any reason to suspect (due to time since last use or the size of the container) that a container of a PECs may have developed pressure, release the pressure by unscrewing the cap, using protective heavy-duty gloves, chemically resistant apron, safety glasses, face shield, and a safety shield between you and the container.

Make sure everyone who uses chemicals that are explosive or could become potentially explosive are thoroughly trained in safe storage methods, conditions to avoid (e.g., contamination), the hazards of the chemical, and disposal procedures.

Chemically reactive substances are stored in designated cabinets in secondary containment and segregated from incompatibles.

A1.9.6   Spill and Incident Procedures

Assess the extent of danger. Help contaminated or injured persons. Evacuate the spill area. Avoid breathing vapors. If possible, confine the spill to a small area using a spill kit or absorbent material. Keep others from entering contaminated area (e.g., caution tape, barriers, etc.).

Small (<500 mL)—If you have training and do not perceive the risk to be greater than normal laboratory operations, use appropriate PPE and cleanup materials for the chemical spilled. Double bag spill waste in clear plastic bags, label, and arrange for chemical waste pickup.

Large (>500 mL)—Call for help. Notify others in area of spill. Turn off ignition sources in area. Evacuate area and post doors to spill area. Remain on the scene, but at a safe distance, to receive and direct safety personnel when they arrive.

Chemical spill on body or clothes—Remove clothing and rinse body or affected skin with plenty of water or thoroughly in emergency shower for at least 15 min. Seek medical attention.

Chemical splash into eyes—Immediately rinse eyeball and inner surface of eyelid with water from the emergency eyewash station for 15 min by forcibly holding the eye open. Seek medical attention.

A1.9.7   Decontamination

Wearing proper PPE, decontaminate work space, equipment, and bench tops using with 70–75% ethanol. Wash hands and arms with soap and water after finished.

Dispose of all used contaminated disposables as hazardous waste following the Waste Disposal Section.

A1.9.8   Waste Disposal

Waste containers containing PECs must be labeled: "EXPLOSION RISK." All waste must be disposed of as hazardous waste by trained lab workers. General disposal guidelines: affix to all waste containers a hazardous waste tag; store hazardous waste in closed containers, in secondary containment, and in a designated location; do not let waste enter the building drains, as discharge into the environment must be prevented; double-bag dry waste using transparent bags; waste must be under the control of the person/laboratory during its generation and until disposal; dispose of routinely generated chemical waste within 90 days.

A1.9.9   Prior Approval/Review Required

All work with PECs require the following prior to beginning work: documented specific training on the techniques and processes to be used; read and understand the relevant Safety Data Sheet; demonstrate competence to perform work; review this document when there are any changes to procedures, personnel, equipment, or when an incident or near miss occurs.

A1.9.10   Designated Area

Work should be completed in a laboratory fume hood. All hoods in our lab are designated for PECs.

- All chemicals containing PECs must be secondarily contained.
- All PECs need to be labeled as "EXPLOSION RISK."
- Signage is required for the container, designated work area, and storage location. Sign wording must state the following: "EXPLOSION RISK."

A1.9.11   Safety Data Sheets

SDSs can be found online.

A1.9.12   Detailed Protocol

When working in the lab, a laboratory worker must be cognizant of all of the SDS and safety information presented in this document; and must find/follow a literature experimental procedure describing the use of this reagent in a related chemical transformation. If a pertinent literature protocol cannot be found, the researcher must discuss the planned experiment with the laboratory supervisor prior to using this reagent; not deviate from the literature experimental protocol mentioned in either temperature or pressure without prior approval from the supervisor; follow all related procedures in the laboratory (PPE, syringe techniques, waste disposal, etc. as appropriately modified by any specific information in the SDS information presented in this document); employ <0.1 mol of this PEC in any given reaction (larger quantities require the approval of supervisor), and discuss all issues or concerns regarding PECs with the laboratory supervisor prior to its use.

A1.10 Acutely Toxic Chemicals

A1.10.1 Hazard Overview

For a large number of toxic chemicals, it is important to recognize the hazards associated with the transportation, operation, and storage of these chemicals. An acutely toxic chemical is considered a chemical falling within any of the following categories: a chemical with a median lethal dose (LD50) of 50 mg or less per kg of body weight when administered orally to albino rats weighing between 200 and 300 gm each; a chemical with a median lethal dose (LD50) of 200 mg or less per kg of body weight when administered by continuous contact for 24 h (or less if death occurs within 24 h) with the bare skin of albino rabbits weighing between 2 and 3 Kg each; a chemical that has a median lethal concentration (LC50) in air of 5000 ppm by volume or less of gas or vapor, or 50 mg per liter or less of mist, fume, or dust, when administered by continuous inhalation for 1 h (or less if death occurs within 1 h) to albino rats weighing between 200 and 300 gm each.

A1.10.2 Hazardous Chemical(s)

This list is provided as a guide and is not all inclusive. Review safety data sheets.

These compounds include acrolein, acryloyl chloride, benzyl chloride, bromine, chlorine dioxide, chlorine trifluoride, chloropicrin, cyanogen chloride, cyanuric fluoride, decaborane, dimethylsulfate, ethylene chlorohydrin, ethylene fluorohydrin, hexamethylene diisocyanate, iron pentacarbonyl, methyl acrylonitrile, methyl chloroformate, methyl fluorosulfate, methyl fluoroacetate, methyl hydrazine, methyl mercury, methyltrichlorosilane, methyl vinyl ketone, nickel carbonyl, nitrogen tetroxide, nitrogen trioxide, methyl tin compounds, osmium tetroxide, ozone, pentaborane, phosphorus oxychloride, phosphorus trichloride, sulfur monochloride, sulfur pentafluoride, sulfuryl chloride, tetramethyl succinonitrile, tetranitromethane, thionyl chloride, toluene-2,4-diisocyanate, trichloromethane sulfenyl chloride, trichloro(chloromethyl)silane.

A1.10.3 Personal Protective Equipment

A1.10.3.1 Eye Protection

At a minimum, approved safety glasses with side shields should be worn. Chemical splash goggles should be worn when working with larger quantities or concentrated aqueous solutions of acids.

A1.10.3.2   Skin and Body Protection

Wear a chemical-resistant lab coat, long pants, and closed-toe shoes. These laboratory coats must be appropriately sized for the individual and be buttoned to their full length. Laboratory coat sleeves must be of a sufficient length to prevent skin exposure while wearing gloves. A chemical-resistant apron should be used when transferring or using large quantities of acids or concentrated aqueous solutions, when acid splashing is a possibility.

A1.10.3.3   Hand Protection

At a minimum, wear a nitrile chemical-resistant glove. Consult with your preferred glove manufacturer to ensure that the gloves you plan on using are compatible with the chemical and usage.

A1.10.4   Engineering/Ventilation Controls

All chemicals should be transferred and used in a laboratory chemical fume hood with the sash at the certified position or lower. The hood flow alarm should be checked to be operating correctly prior to using the hood. Use HEPA filters, carbon filters, or scrubber systems with containment devices to protect effluent and vacuum lines, pumps, and the environment whenever feasible. Use ventilated containment to weigh out solid chemicals. Alternatively, the tare method can be used to prevent inhalation of the chemical. While working in a laboratory hood, the chemical is added to a pre-weighed container. The container is then sealed and can be reweighed outside the hood. If chemical needs to be added or removed, this manipulation is carried out in the hood. In this manner, all open chemical handling is conducted in the laboratory hood.

A1.10.5   Special Handling Procedures and Storage Requirements

It is important that chemical users track and dispose of chemicals before they become a problem. Proper inventory management systems can help mitigate risk to personnel and avert disposal costs.

A1.10.6   Spill and Incident Procedures

Assess the extent of danger. Help contaminated or injured persons. Evacuate the spill area. Avoid breathing vapors. If possible, confine the spill to a small area using a spill kit or absorbent material. Keep others from entering contaminated area (e.g., caution tape, barriers, etc.).

Small (<500 mL)—If you have training and do not perceive the risk to be greater than normal laboratory operations, use appropriate PPE and cleanup materials for the chemical spilled. Double bag spill waste in clear plastic bags, label, and arrange for chemical waste pickup.

Large (>500 mL)—Call for help. Notify others in area of spill. Turn off ignition sources in area. Evacuate area and post doors to spill area. Remain on the scene, but at a safe distance, to receive and direct safety personnel when they arrive.

Chemical spill on body or clothes—Remove clothing and rinse body or affected skin with plenty of water or thoroughly in emergency shower for at least 15 min. Seek medical attention.

Chemical splash into eyes—Immediately rinse eyeball and inner surface of eyelid with water from the emergency eyewash station for 15 min by forcibly holding the eye open. Seek medical attention.

A1.10.7   Decontamination

Wearing proper PPE, decontaminate work space, equipment, and bench tops using with 70–75% ethanol. Wash hands and arms with soap and water after finished. Dispose of all used contaminated disposables as hazardous waste following the Waste Disposal Section.

A1.10.8   Waste Disposal

All waste must be disposed through the hazardous waste program. Staff dealing with hazardous waste disposal should have completed waste management training. Label all waste with the chemical contents and the appropriate hazard warning. Store hazardous waste in closed containers, in secondary containment, and in a designated location. Do not let waste enter the building drains, as discharge into the environment must be prevented. Double-bag dry waste using transparent bags. Waste must be under the control of the person generating and disposing of it. Dispose of routinely generated chemical waste within 90 days.

A1.10.9   Prior Approval/Review Required

All work with acutely toxic chemical must be preapproved by the laboratory supervisor prior to use and all training must be well documented.

A1.10.10   Designated Area

Work should be completed in a laboratory fume hood. All hoods in our lab are designated for acutely toxic chemicals. Designated areas must be clearly marked with signs that identify the chemical hazard and include an appropriate warning; for example: WARNING! ACUTELY TOXIC MATERIAL WORK AREA!

A1.10.11   Safety Data Sheets

SDSs can be found online.

A1.10.12   Detailed Protocol

Ozone: The toxicity of ozone to the human respiratory system demands that ozone generators be used in an efficient fume hood. The ozone-containing gas stream is slowly bubbled into a solution of the reactant (in methanol, acetic acid, chloroform, hexanes, or ethyl acetate) to conduct ozonation. The rate of ozone production is defined by the input gas composition, its flow rate, and the electrical power, but is c. 5 mmol/h with air and 0.5 mol/h with O_2.

Halides: Many of the compounds in this hazard class are reactive alkylating/acylating agents. Any other compounds in those structural classes (acyl bromides, acyl fluorides, sulfonyl halides, vinyl carbonyl compounds) likely pose similar hazards and should be treated in the same way as the explicitly listed examples. A main concern is how to handle residual reagent in reactions that do not go to completion or that require an excess. Quenching of these reagent types can be accomplished with concentrated aqueous ammonia solution, which should give product(s) that are substituted amines and not hazardous. The reaction of a significant quantity of halide with ammonia can be violent, so this step must be performed with caution. Glassware can also be carefully rinsed with aqueous ammonia to decontaminate it (prior to conventional dishwashing practices), and a flask of aqueous ammonia solution can be kept nearby the reaction to deal with any spills that might occur.

The research laboratory requires variation in reaction conditions to develop and optimize new chemical transformations. The researcher must seek literature precedent for reaction conditions that have reasonable similarities to new chemistry that is planned with acutely toxic chemicals described in this protocol. The researcher must also consult the laboratory supervisor or a designated, experienced research coworker for approval to proceed with chemical transformations that have little literature or local research group precedent. Laboratory supervisor approval must also be obtained for fivefold scale-up of new chemistry transformations.

A1.11   Acutely Toxic Gases

A1.11.1   Hazard Overview

For a large number of toxic compressed gases, it is important to recognize the hazards associated with the transportation, operation, and storage of these gases.

Toxic gases are gases that may cause significant acute health effects at low concentrations. Health effects may include severe skin or eye irritation, pulmonary edema, neurotoxicity, or other potentially fatal conditions. The criteria used to establish the list are as follows: (1) An NFPA health rating of 3 or 4, (2) An NFPA health rating of 2 with poor physiological warning properties, (3) Pyrophoric (self-igniting) characteristics, or (4) Extremely low occupational exposure limits in the absence of an NFPA health rating.

A1.11.2   Hazardous Chemical(s) or Class of Hazardous Chemical(s)

These include anhydrous ammonia, arsine, carbon monoxide, chlorine, dichlorosilane, fluorine, hydrogen chloride, hydrogen bromide, hydrogen sulfide, nitric oxide, phosgene, phosphine, vinyl chloride.

A1.11.3 Personal Protective Equipment

A1.11.3.1 Eye Protection

At a minimum, approved safety glasses with side shields should be worn. Chemical splash goggles should be worn when working with larger quantities or concentrated aqueous solutions of acids.

A1.11.3.2 Skin and Body Protection

Wear a chemical-resistant lab coat, long pants, and closed-toe shoes. These laboratory coats must be appropriately sized for the individual and be buttoned to their full length. Laboratory coat sleeves must be of a sufficient length to prevent skin exposure while wearing gloves. A chemical-resistant apron should be used when transferring or using large quantities of acids or concentrated aqueous solutions, when acid splashing is a possibility. If working with a pyrophoric gas (e.g., phosphine), a flame-resistant lab coat is required.

A1.11.3.3 Hand Protection

At a minimum, wear a nitrile chemical-resistant glove. Consult with your preferred glove manufacturer to ensure that the gloves you plan on using are compatible with the chemical and usage.

A1.11.4 Engineering/Ventilation Controls

A ventilation monitor is required on each lab hood or gas cabinet in which toxic gases are used and stored. Acceptable monitors include a magnehelic gauge, audible and visual alarms, inclined manometer, or other devices that indicate the enclosure is actively ventilated. Manometers and gauges should be clearly marked to indicate safe pressure limits.

Electronic toxic gas monitors with alarms should be installed and continuously operated wherever a toxic gas is used that has a high concentration, large quantity, and/or poor physiological warning properties (odor or irritation). The requirement for a monitoring system will be decided on a case-by-case basis and will be required more commonly for continuous operations and long-term research situations. A toxic gas has poor warning properties when such properties are only noticeable at or above harmful concentrations (e.g., the PEL). Some toxic gases have poor warning properties. Gas monitoring equipment must be able to detect concentrations at or below the PEL.

Toxic gas monitors and alarms should be connected to an emergency power source. In the event of a power failure, the detection system should continue to operate without interruption, or gas systems should automatically shut down at the source. Power connections, control switches, and adjustments that affect the detection system operation should be protected from direct access by locks on the enclosures.

All gas-monitoring systems should have audible and visible alarms in the following locations: gas supply location, gas use or operator room, and outside the gas use room (e.g., corridor); an alarm status and gas concentration readout panel located outside the gas use room; local audible and visual alarms specific and distinct from fire alarm bells and signs to indicate the alarm's meaning and required personnel action; the toxic gas alarm level set point set at the PEL or threshold limit value.

A1.11.5 Special Handling Procedures and Storage Requirements

All transport of toxic gases must be conducted as follows: gas cylinders must be secured to the transport (cart, hand truck, etc.); cylinders must be continuously attended during transport; cylinders must be clearly labeled with content and hazard information; cylinder caps must be in place. These requirements apply to all listed toxic gas containers, including empty and partially full cylinders.

Upon receipt of toxic gases, cylinders shall be temporarily stored in a well-ventilated area that is attended or locked at all times. All cylinders shall be immediately leak tested with a leak indicating solution and must be clearly labeled with content and hazard information. Temporary storage locations shall have appropriate signage in place. Cylinders must be seismically secured at all locations with chains at two contact points on the cylinder body, using unistruts or an equivalent. Seismic securing should prevent cylinders from rolling, shifting, or falling.

Laboratory storage of all toxic gas cylinders shall be in a mechanically ventilated, lockable area. Examples of mechanical ventilation include vented gas cabinets and fume hoods. Rooms containing toxic gases shall be locked when not occupied by authorized persons. All cylinders and gas cabinets must be clearly labeled with content and hazard information. Cylinders shall be seismically secured at all locations with chains (2 contact points), using unistruts or an equivalent for cylinders larger than lecture bottles. Lecture bottles must be secured to a stable surface.

All regulators, valves, and lines must be chemically compatible with the gases being used. Compatibility can be determined by contacting the gas vendor. Regulator/line systems must be leak tested immediately after assembly and before each use. Regulators shall be compatible with the size and type of gas cylinder being used, and rated for full cylinder pressure.

All toxic gas cylinders and reaction vessels/chambers shall be kept in ventilated fume hoods during use and storage. Air-flow velocities at all openings in the hoods must be 0.5 m/s (100 fpm) or greater while in the open position. Where regular access is needed, small access doors must be used to minimize exhaust flow reduction.

All lines or ducts carrying purged or exhausted emissions of toxic gases must be connected to a mechanical exhaust system that discharges to a safe location (i.e., presents no potential for reentrainment into any building supply air intake or occupied area).

Significant emissions of corrosive or toxic gases require an emission control device (e.g., scrubber, flare device, adsorbent) before the purged gas can be vented into the exhaust duct system. Significant emissions are defined as duct concentrations that result in duct corrosion or acute health risk to persons exposed near exhaust fan stacks as determined by release modeling.

A1.11.6   Spill and Incident Procedures

Emergency procedure for leaking gas cylinders—http://www.airproducts.com/~/media/ Files/PDF/company/safetygram-11.pdf.

A1.11.7   Decontamination

Not relevant, as gases will diffuse away. In the event of accidental release, leave the laboratory and prevent others from entering.

A1.11.8   Waste Disposal

All empty toxic gas cylinders shall be labeled as empty. Depleted toxic gas cylinders should be returnable according to vendor guidelines. The purchase of any gases that will not be completely used in the course of research must be approved by the vendor for return. Disposal of toxic gas cylinders, even when empty, may entail extraordinary costs.

A1.11.9   Prior Approval/Review Required

All work with acutely toxic gases must be preapproved by the laboratory supervisor prior to use and all training must be well documented.

A1.11.10   Designated Area

A designated area shall be established where limited access, special procedures, knowledge, and work skills are required. All work should be performed in a laboratory fume hood. All hoods in our lab are designated for work with toxic gases. When being used for this purpose, they must be clearly marked with signs that identify the chemical hazard and include an appropriate warning; for example: WARNING! ACUTELY TOXIC GAS WORK AREA!

A1.11.11   Safety Data Sheets

SDSs can be found online.

A1.11.12   Detailed Protocol

Anhydrous HCl in methanol should not be stored for long periods because it is so corrosive, but it is perfectly reasonable to prepare such a solution and keep it in the hood for several days. It is prepared by carefully bubbling gaseous HCl from a tank into a known volume of dry methanol in an ice-cooled, tared flask (caution: HCl has a quite substantial heat of solution, so this is a very exothermic process. It is important to very carefully lower the sparging tube into the solution to avoid drawing methanol back into the gas line. The flask may also be cooled in an ice bath). After the reaction has returned to ambient temperature, the flask is weighed and the HCl concentration is determined.

Lab workers using acutely toxic gases must demonstrate competence to the laboratory supervisor by being able to (1) identify the hazards and list any particularly hazardous handling techniques (use of a Schlenck line, canula transfer, extremes of pressure or temperature, etc.), (2) list the foreseeable emergency situations, (3) describe the proper response to the emergency situations, and (4) know the control measures to minimize the risks.

Appendix 2
Synthetic Solvent Selection Chart

The chart on the following page (Fig. A2.1) is based on principal components analysis of the properties mp, bp, dielectric constant (ε); dipole moment (μ); refractive index (η); ET (spectroscopically determined effect on a solvatochromatic dye); and *log P* for >80 solvents (Carlson et al., 1985). The eigenvectors were projected into two dimensions, t_1 and t_2, representing a conflation of properties that most differentiate these solvents, accounting for around 80% of their variance. Polarity correlates with t_1, while polarizability correlates with t_2. Solvents in the vicinity of one another in this chart should have similar properties.

This chart can be used in several ways. For reactions with a known solvent dependence, select solvents in the vicinity of the known best solvent(s). For reactions with an unknown solvent dependence:

- select solvents that give uniform coverage of the solvent space (ie, a regular lattice with a density consistent with the number of experiments desired). Both aprotic and protic solvents might be selected when compatible with reaction conditions.
- select solvents as dissimilar as possible. A D-optimal design selects solvents at the extreme periphery of the solvent space. A quadratic D-optimal design also includes a solvent near the origin.
- select typical solvents from each class (polar/nonpolar, protic/aprotic, etc.).

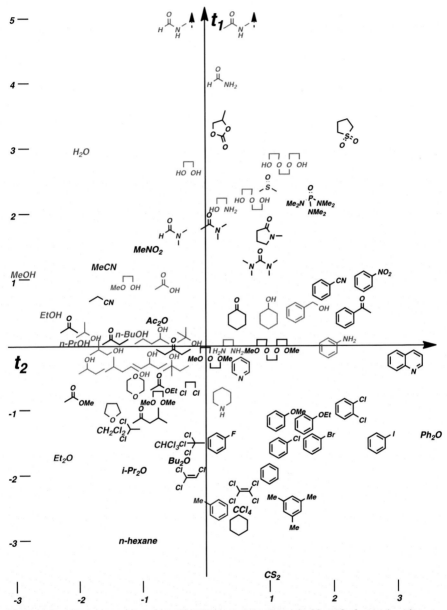

Figure A2.1 Solvents known to swell polystyrene are shown in *blue (dark gray in print version)*. Protic solvents are shown in *red (light gray in print version)*. Aprotic solvents are shown in *black*. *N*-methylformamide and *N*-methylacetamide are shown slightly lower than their t_1 coordinates, 5.4 and 5.3, to compact the chart.

Reference

Carlson, R., Lunstedt, T., Albano, C., 1985. Screening of suitable solvents in organic synthesis. Strategies for solvent selection. Acta Chem. Scand. B 39, 79–91. http://dx.doi.org/ 10.3891/acta.chem.scand.39b-0079.

Appendix 3
Solvent Miscibility

Solvents are only considered miscible when they mix in all proportions. Immiscible means two solvents that produce two phases when mixed in some proportions. The miscibility of solvents commonly used in organic synthesis is presented in the chart on the following page. It makes apparent the organic solvents that are specifically immiscible, which can be a great help in solvent partitioning work-ups of reactions that involve polar, nonvolatile solvents. By judicious choice of an organic phase, these can be driven to the aqueous phase where they can be easily removed. Miscibility is also crucial for solvent mixtures used in chromatography (Fig. A3.1).

Solvent Miscibility and Solubility

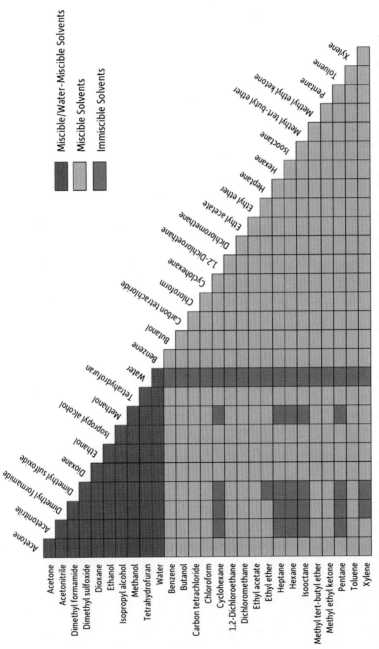

Figure A3.1 The miscibility of 27 solvents. The three classifications are miscible/water–miscible solvents (*blue, gray in print version*)—polar organics 100% miscible with water and other polar organics. Miscible solvents (*yellow, light gray in print version*)—organic solvents with limited to no miscibility with water or the other polar organics in *blue*. Solvents in *yellow* are generally 100% miscible with one another. Immiscible solvents (*purple, dark gray in print version*)—organic solvents that are immiscible with water to a high degree, as well as to some polar organics. The purple row of cells below water constitutes a line of demarcation. Any solvent to its right is miscible with any other solvent in this region. Solvents to its left are of intermediate polarity. This chart was developed by Vernon Bartlett of Restek. Reproduced by permission of Restek Corporation (©2015 by Restek Corporation).

Appendix 4
Freezing Points of Common Organic Solvents

Table A4.1 Freezing Points of Organic Solvents/Reagents (°C)

Sulfolane	28	1,2-Dichloroethane	−36
Cyclohexanol	26	3-Pentanone	−39
2-Methyl-2-propanol	26	Fluorobenzene	−42
Dimethyl sulfoxide	18	Pyridine	−42
Glycerol	18	Acetonitrile	−44
Acetic acid	17	2,2,2-Trifluoroethanol	−44
p-Xylene	13	1,3,5-Trimethylbenzene	−45
1,4-Dioxane	12	Chlorobenzene	−46
1,2-Dibromoethane	10	Propylene carbonate	−49
Formic acid	8	N,N-dimethylformamide	−60
Hexamethylphosphoramide	7	Diisopropylamine	−61
Cyclohexane	7	Chloroform	−64
Benzene	5	Diglyme	−68
N,N-dimethylaniline	2	1,2-Dimethoxyethane	−68
Hydrazine	1	Ethyl acetate	−84
Water	0	2-Methoxyethanol	−85
Tetramethylurea	−1	Butanone	−87
2,6-Dimethylpyridine	−6	1-Butanol	−90
Bromine	−7	Isopropyl alcohol	−90
2-Methyl-2-butanol	−9	Perfluorohexane (FC-72)	−90
Piperidine	−11	Heptane	−91
Quinoline	−15	Cyclopentane	−93
1-Octanol	−15	Acetone	−95
Trifluoroacetic acid	−15	Dibutyl ether	−95
o-Dichlorobenzene	−17	Dichloromethane	−95
N,N-dimethylacetamide	−19	Hexane	−95
Dimethyl propylene urea	−20	Toluene	−95
Carbon tetrachloride	−22	1,1-Dichloroethane	−97
Tetrachloroethylene	−22	Methanol	−98
N-methyl-2-pyrrolidone	−23	Methyl acetate	−98
o-Xylene	−25	Dimethoxymethane	−105
Nitromethane	−28	Methyl tert-butyl ether	−109
Benzotrifluoride	−29	Tetrahydrofuran	−109
1,1,1-Trichloroethane	−30	Diethyl ether	−117
Bromobenzene	−31	Pentane	−130

Appendix 5
Toxicities of Common Organic Solvents

Solvent toxicity is an important consideration because the quantities of solvents used in the synthesis lab far exceed those of reagents. The routes of exposure to be concerned with are skin and inhalation, which are of course related to the controls used to minimize the hazards: personal protective equipment and fume hoods. Threshold limit values (TLVs) are a measure of inhalation toxicity under chronic exposure. They are expressed as parts per million by volume. The TLV is the level to which someone could be exposed to a solvent throughout a normal work week without adverse effects. The TLVs of a selection of common solvents are provided in Table A5.1. Of course, no one using a solvent in a research lab would be exposed in this way, but relative TLVs still give some indication of the risk of various solvents. However, there is also an interplay of solvent volatility with its intrinsic inhalation hazard. That is, a solvent could be toxic but nonvolatile and pose less practical risk than a solvent that is objectively less toxic but much more volatile. Of course, volatile organic solvents are typically valued for that property, which enables their ready evaporation to give desired products.

The need to evaluate these two features of solvents led to the development of the vapor hazard ratio (VHR), essentially the vapor pressure (concentration in ppm at saturation) divided by the exposure limit. One must take care in comparing VHRs from different sources that may be based on different exposure limits (OEL, TLV, etc.). The range of values for VHRs is sufficiently large that some tabulations include the logarithm of the VHR, called the vapor hazard index (VHI). Two different compilations of VHRs are provided in Tables A5.2 and A5.3.

A feature of note in Table A5.2 is a VHR for hexane that is greater than expected compared to homologous n-alkanes. This is related to its widely discussed toxicity based on oxidative metabolism to 2,5-hexanedione. Also included in this table is 2-hexanone, which can also be converted to 2,5-hexanedione. Solvent hexane used in the laboratory is most often hexanes, meaning a mixture of C6 isomers that includes methylcyclopentane.

Table A5.1 **Inhalation Toxicity of 25 Common Organic Solvents**

Solvent	Threshold Limit Value (ppm)
Formic acid	5
Acetic acid	10
Chlorobenzene	10
1-Butanol	20
Ammonia	25
tert-Butyl methyl ether	50
Dichloromethane	50
iso-Butanol	50
3-Methyl-1-butanol	100
2-Butanol	100
tert-Butanol	100
Cyclohexane	100
1,1-Dichloroethane	100
iso-Propanol	200
1,2-Dichloroethylene	200
Methyl acetate	200
3-Pentanone	200
n-Octane	300
Diethyl ether	400
Ethyl acetate	400
n-Heptane	400
Methylcyclohexane	400
Acetone	500
Pentane	600
Dimethoxymethane	1000

Data from Reichardt, C., Welton, T., 2010. Solvents and Solvent Effects in Organic Chemistry, fourth ed. Wiley-VCH, Weinheim.

Table A5.2 **Vapor Hazard Ratio (VHR) and Vapor Hazard Index (VHI) of 10 Organic Compounds**

Compound	VHR	VHI
Benzene	251,000	5.4
Carbon disulfide	47,500	4.7
Carbon tetrachloride	30,000	4.5
2-Hexanone	3000	3.5
Cyclohexane	1300	3.1
n-Pentane	1100	3.1
n-Hexane	4000	3.6
n-Heptane	150	2.2
n-Octane	61	1.8
n-Nonane	30	1.4

Data from Smith, P.A., 2008. Toxicology of organic solvents. In: Luttrell, W.E., Jederberg, W.W., Still, K.R. (Eds.), Toxicology Principles for the Industrial Hygienist. Amer. Industrial Hygiene Assn., Fairfax, pp. 186–208.

Table A5.3 **Vapor Hazard Ratio (VHR) of 32 Common Organic Solvents**

Solvent	VHR
Chloroform	100,000
Benzene	99,000
Carbon tetrachloride	60,000
Triethylamine	36,000
Chlorobenzene	12,000
tert-Butyl methyl ether	11,000
Formic acid	8800
n-Hexane	8000
Diethyl ether	5900
Dichloromethane	5800
Tetrahydrofuran	4000
Pyridine	4200
Acetonitrile	2400
2-Methoxyethanol	1600
1,2-Dichloroethylene	1500
1,4-Dioxane	1440
Methyl acetate	1200
Cyclohexane	1000
Methanol	650
tert-Butanol	600
Toluene	580
Acetone	500
Ethyl acetate	490
Dimethoxymethane	440
Dimethylformamide	390
Nitromethane	370
3-Pentanone	180
2-Butanol	170
iso-Propanol	110
n-Heptane	94
p-Dichlorobenzene	92
Ethanol	59

Data from Pitt, M.J., 2006. A vapor hazard index for use with COSHH and DSEAR. Hazards XIX – Process Safety and Environmental Protection, IChemE, pp. 95–107.

Appendix 6
Recipes for TLC Stains

Staining solutions are ideally stored in 100-mL-wide mouth jars covered with aluminum foil, a watch glass, or a screw cap to minimize evaporation.

A6.1 General Stains

Phosphomolybdic acid (PMA): Dissolve 10 g of PMA in 100 mL of absolute ethanol.

Potassium permanganate: Dissolve 1.5 g of $KMnO_4$, 10 g of K_2CO_3, and 1.25 mL of 10% NaOH in 200 mL of water.

Vanillin: Dissolve 15 g of vanillin in 250 mL of ethanol and add 2.5 mL of concentrated sulfuric acid.

p-Anisaldehyde A: To 135 mL of absolute ethanol, add 5 mL of concentrated sulfuric acid and 1.5 mL of glacial acetic acid. Allow the solution to cool to room temperature. Add 3.7 mL of p-anisaldehyde. Stir the solution vigorously to ensure homogeneity. Store refrigerated.

p-Anisaldehyde B: Dissolve 2.5 mL of concentrated sulfuric acid and 15 mL of p-anisaldehyde in 250 mL of 95% ethanol.

Cerium molybdate (Hanessian's Stain): To 235 mL of distilled water, add 12 g of ammonium molybdate, 0.5 g of ceric ammonium molybdate, and 15 mL of concentrated sulfuric acid. Wrap the jar with aluminum foil as the stain may be somewhat photosensitive.

A6.2 Specialized Stains

Dinitrophenylhydrazine (DNP): Dissolve 12 g of 2,4-DNP, 60 mL of concentrated sulfuric acid, and 80 mL of water in 200 mL of 95% ethanol.

Ninhydrin: Dissolve 0.3–1.5 g of ninhydrin in 100 mL of n-butanol. Then add 3.0 mL of acetic acid.

Bromocresol green: Dissolve 0.04 g of bromocresol green in 100 mL of absolute ethanol, which should give a colorless solution. Add a 0.1 M solution of aqueous NaOH dropwise until a blue color just appears.

Ceric ammonium sulfate: Prepare a 1% (w/v) solution of cerium (IV) ammonium sulfate in 50% phosphoric acid.

Ceric ammonium molybdate: Dissolve 0.5 g of ceric ammonium sulfate (Ce $(NH_4)_4(SO_4)_4 \bullet 2H_2O$) and 12 g of ammonium molybdate $((NH_4)_6Mo_7O_{24} \bullet 4H_2O)$ in 235 mL of water. Add 15 mL of concentrated sulfuric acid.

Cerium sulfate: Prepare a 10% (w/v) solution of cerium (IV) sulfate in 15% sulfuric acid.

Ferric chloride: Prepare a 1% (w/v) solution of 1% ferric (III) chloride in 50% aqueous methanol.

Ehrlich's reagent: Dissolve 1.0 g of p-dimethylaminobenzaldehyde in 75 mL of MeOH and add 50 mL of concentrated HCl.

Dragendorff–Munier stain: Dissolve 10 g of KI in 40 mL water. Add 1.5 g of $Bi(NO_3)_3$ and 20 g of tartaric acid in 80 mL of water. Stir for 15 min and then filter.

Appendix 7
NMR Spectral Data of Common Contaminants of Organic and Organometallic Reaction Products

The tabulated data are broken into several different tables because they come from multiple sources (Tables A7.1–A7.6).

Table A7.1 ^{1}H NMR Data of Common Solvents and Contaminants

	Proton	Mult	$CDCl_3$	$(CD_3)_2CO$	$(CD_3)_2SO$	C_6D_6	CD_3CN	CD_3OD	D_2O
Solvent residual signals			7.26	2.05	2.50	7.16	1.94	3.31	4.79
H_2O		s	1.56	2.84	3.33	0.40	2.13	4.87	ND
Acetic acid	CH_3	s	2.10	1.96	1.91	1.55	1.96	1.99	2.08
Acetonitrile	CH_3	s	2.10	2.05	2.07	1.55	1.96	2.03	2.06
Benzene	CH	s	7.36	7.36	7.37	7.15	7.37	7.33	
tert-Butyl alcohol	CH_3	s	1.28	1.18	1.11	1.05	1.16	1.40	1.24
	OH	s		4.19	1.11	1.55	2.18	ND	ND
tert-Butyl methyl ether	CH_3	s	1.19	1.13	1.11	1.07	1.14	1.15	1.21
Chloroform		s	7.26	8.02	8.32	6.15	7.58	7.90	ND
Cyclohexane	CH_2	s	1.43	1.43	1.40	1.40	1.44	1.45	ND
1,2-Dichloroethane	CH_2	s	3.73	3.87	3.90	2.90	3.81	3.78	ND
Dichloromethane	CH_2	s	5.30	5.63	5.76	4.27	5.44	5.49	ND
Diethyl ether	CH_3	t	1.21	1.11	1.09	1.11	1.12	1.18	1.17
	CH_2	q	3.48	3.41	3.38	3.26	3.42	3.49	3.56
Diglyme	CH_2	m	3.65	3.56	3.51	3.46	3.53	3.61	3.67
	CH_2	m	3.57	3.47	3.38	3.34	3.45	3.58	3.61
	OCH_3	s	3.39	3.28	3.24	3.11	3.29	3.35	3.37
1,2-Dimethoxyethane	CH_3	s	3.40	3.28	3.24	3.12	3.28	3.35	3.37
	CH_2	s	3.55	3.46	3.43	3.33	3.45	3.52	3.60
Dimethylacetamide	CH_3CO	s	2.09	1.97	1.96	1.60	1.97	2.07	2.08
	NCH_3	s	3.02	3.00	2.94	2.57	2.96	3.31	3.06
	NCH_3	s	2.94	2.83	2.78	2.05	2.83	2.92	2.90
Dimethylformamide	CH	s	8.02	7.96	7.95	7.63	7.92	7.97	7.92
	CH_3	s	2.96	2.94	2.89	2.36	2.89	2.99	3.01
	CH_3	s	2.88	2.78	2.73	1.86	2.77	2.86	2.85
Dimethyl sulfoxide	CH_3	s	2.62	2.52	2.54	1.68	2.50	2.65	2.71
Dioxane	CH_2	s	3.71	3.59	3.57	3.35	3.60	3.66	3.75
Ethanol	CH_3	t	1.25	1.12	1.06	0.96	1.12	1.19	1.17

Compound	Proton	Mult							
	CH₂	q	3.72	3.57	3.44	3.34	3.54	3.60	3.65
	OH	s	1.32	3.39	4.63		2.47	ND	ND
Ethyl acetate	CH₃CO	s	2.05	1.97	1.99	1.65	1.97	2.01	2.07
	CH₂CH₃	q	4.12	4.05	4.03	3.89	4.06	4.09	4.14
	CH₂CH₃	t	1.26	1.20	1.17	0.92	1.20	1.24	1.24
Ethyl methyl ketone	CH₃CO	s	2.14	2.07	2.07	1.58	2.06	2.12	2.19
	CH₂CH₃	q	2.46	2.45	2.43	1.81	2.43	2.50	3.18
	CH₂CH₃	t	1.06	0.96	0.91	0.85	0.96	1.01	1.26
Ethylene glycol	CH	s	3.76	3.28	3.34	3.41	3.51	3.59	3.65
"Grease"	CH₃	m	0.86	0.87		0.92	0.86	0.88	ND
	CH₂	br s	1.26	1.29		1.36	1.27	1.29	ND
n-Hexane	CH₃	t	0.88	0.88	0.86	0.89	0.89	0.90	ND
	CH₂	m	1.26	1.28	1.25	1.24	1.28	1.29	ND
HMPA	CH₃	d	2.65	2.59	2.53	2.40	2.57	2.64	2.61
Methanol	CH₃	s	3.49	3.31	3.16	3.07	3.28	3.34	3.34
	OH	s	1.09	3.12	4.01		2.16		ND
Nitromethane	CH₃	s	4.33	4.43	4.42	2.94	4.31	4.34	4.40
n-Pentane	CH₃	t	0.88	0.88	0.86	0.87	0.89	0.90	ND
	CH₂	m	1.27	1.27	1.27	1.23	1.29	1.29	ND
2-Propanol	CH₃	d	1.22	1.10	1.04	0.95	1.09	1.50	1.17
	CH	sep	4.04	3.90	3.78	3.67	3.87	3.92	4.02
Pyridine	CH(2)	m	8.62	8.58	8.58	8.53	8.57	8.53	8.52
	CH(3)	m	7.29	7.35	7.39	6.66	7.33	7.44	7.45
	CH(4)	m	7.68	7.76	7.79	6.98	7.73	7.85	7.87
Silicone grease	CH₃	s	0.07	0.13		0.29	0.08	0.10	ND
Tetrahydrofuran	CH₂	m	1.85	1.79	1.76	1.40	1.80	1.87	1.88
	CH₂O	m	3.76	3.63	3.60	3.57	3.64	3.71	3.74
Toluene	CH₃	s	2.36	2.32	2.30	2.11	2.33	2.32	ND
	CH(o/p)	m	7.17	7.1–7.2	7.18	7.02	7.1–7.3	7.16	ND
	CH(m)	m	7.25	7.1–7.2	7.25	7.13	7.1–7.3	7.16	ND

Reprinted with Permission from Gottlieb, Kotlyar, Nudelman, 1997. Copyright 1997, American Chemical Society.

Table A7.2 ^{13}C NMR data of Common Solvents and Contaminants

	Carbon	$CDCl_3$	$(CD_3)_2CO$	$(CD_3)_2SO$	C_6D_6	CD_3CN	CD_3OD	D_2O
Solvent signals		77.16 ± 0.06	29.84 ± 0.01, 206.26 ± 0.13	39.52 ± 0.06	128.06 ± 0.02	1.32 ± 0.02, 118.26 ± 0.02	49.00 ± 0.01	
Acetic acid	CO	175.99	172.31	171.93	175.82	173.21	175.11	177.21
	CH_3	20.81	20.51	20.95	20.37	20.73	20.56	21.03
Acetone	CO	207.07	205.87	206.31	204.43	207.43	209.67	215.94
	CH_3	30.92	30.60	30.56	30.14	30.91	30.67	30.89
Acetonitrile	CN	116.43	117.60	117.91	116.02	118.26	118.06	119.68
	CH_3	1.89	1.12	1.03	0.20	1.79	0.85	1.47
Benzene	CH	128.37	129.15	128.30	128.62	129.32	129.34	ND
tert-Butyl alcohol	C	69.15	68.13	66.88	68.19	68.74	69.40	70.36
	CH_3	31.25	30.72	30.38	30.47	30.68	30.91	30.29
tert-Butyl methyl ether	OCH_3	49.45	49.35	48.70	49.19	49.52	49.66	49.37
	C	72.87	72.81	72.04	72.40	73.17	74.32	75.62
Chloroform	CH	77.36	79.19	79.16	77.79	79.17	79.44	ND
Cyclohexane	CH_2	26.94	27.51	26.33	27.23	27.63	27.96	ND
1,2-Dichloroethane	CH_2	43.50	45.25	45.02	43.59	45.54	45.11	ND
Dichloromethane	CH_2	53.52	54.95	54.84	53.46	55.32	54.78	ND
Diethyl ether	CH_3	15.20	15.78	15.12	15.46	15.63	15.46	14.77
	CH_2	65.91	66.12	62.05	65.94	66.32	66.88	66.42
Diglyme	CH_3	59.01	58.77	57.98	58.66	58.90	59.06	58.67
	CH_2	70.51	71.03	69.54	70.87	70.99	71.33	70.05
	CH_2	71.90	72.63	71.25	72.35	72.63	72.92	71.63
1,2-Dimethoxyethane	CH_3	59.08	58.45	58.01	58.68	58.89	59.06	58.67
	CH_2	71.84	72.47	72.07	72.21	72.47	72.72	71.49
Dimethylacetamide	CH_3	21.53	21.51	21.29	21.16	21.76	21.32	21.09
	CO	171.07	170.61	169.54	169.95	171.31	173.32	174.57

Compound	Group							
	NCH_3	35.28	34.89	37.38	34.67	35.17	35.50	35.03
	NCH_3	38.13	37.92	34.42	37.03	38.26	38.43	38.76
Dimethylformamide	CH	162.62	162.79	162.29	162.13	163.31	164.73	165.53
	CH_3	36.50	36.15	35.73	35.25	36.57	36.89	37.54
	CH_3	31.45	31.03	30.73	30.72	31.32	31.61	32.03
Dimethyl sulfoxide	CH_3	40.76	41.23	40.45	40.03	41.31	40.45	39.39
Dioxane	CH_2	67.14	67.60	66.36	67.16	67.72	68.11	67.19
Ethanol	CH_3	18.41	18.89	18.51	18.72	18.80	18.40	17.47
	CH_2	58.28	57.72	56.07	57.86	57.96	58.26	58.05
Ethyl acetate	CH_3CO	21.04	20.83	20.68	20.56	21.16	20.88	21.15
	CO	171.36	170.96	170.31	170.44	171.68	172.89	175.26
	CH_2	60.49	60.56	59.74	60.21	60.98	61.50	62.32
	CH_3	14.19	14.50	14.40	14.19	14.54	14.49	13.92
Ethyl methyl ketone	CH_3CO	29.49	29.30	29.26	28.56	29.60	29.39	29.49
	CO	209.56	208.30	208.72	206.55	209.88	212.16	218.43
	CH_2CH_3	36.89	36.75	35.83	36.36	37.09	37.34	37.27
	CH_2CH_3	7.86	8.03	7.61	7.91	8.14	8.09	7.87
Ethylene glycol	CH_2	63.79	64.26	62.76	64.34	64.22	64.30	63.17
"Grease"	CH_2	29.76	30.73	29.20	30.21	30.86	31.29	ND
n-Hexane	CH_3	14.14	14.34	13.88	14.32	14.43	14.45	ND
	$CH_2(2)$	22.70	23.28	22.05	23.04	23.40	23.68	ND
	$CH_2(3)$	31.64	32.30	30.95	31.96	32.36	32.73	ND
HMPA	CH_3	36.87	37.04	36.42	36.88	37.10	37.00	36.46
Methanol	CH_3	50.41	49.77	48.59	49.97	49.90	49.86	49.50
Nitromethane	CH_3	62.50	63.21	63.28	61.16	63.66	63.08	63.22
n-Pentane	CH_3	14.08	14.29	13.28	14.25	14.37	14.39	ND
	$CH_2(2)$	22.38	22.98	21.70	22.72	23.08	23.38	ND
	$CH_2(3)$	34.16	34.83	33.48	34.45	34.89	35.30	ND
2-Propanol	CH_3	25.14	25.67	25.43	25.18	25.55	25.27	24.38
	CH	64.50	63.85	64.92	64.23	64.30	64.71	64.88
Pyridine	$CH(2)$	149.90	150.67	149.58	150.27	150.76	150.07	149.18

Continued

Table A7.2 ^{13}C NMR Data of Common Solvents and Contaminants—cont'd

	Carbon	CDCl$_3$	(CD$_3$)$_2$CO	(CD$_3$)$_2$SO	C$_6$D$_6$	CD$_3$CN	CD$_3$OD	D$_2$O
	CH(3)	123.75	124.57	123.84	123.58	127.76	125.53	125.12
	CH(4)	135.96	136.56	136.05	135.28	136.89	138.35	138.27
Silicone grease	CH$_3$	1.04	1.40		1.38		2.10	ND
Tetrahydrofuran	CH$_2$	25.62	26.15	25.14	25.72	26.27	26.48	25.67
	CH$_2$O	67.97	68.07	67.03	67.80	68.33	68.83	68.68
Toluene	CH$_3$	21.46	21.46	20.99	21.10	21.50	21.50	ND
	C(i)	137.89	138.48	137.35	137.91	138.90	138.85	ND
	CH(o)	129.07	129.76	128.88	129.33	129.94	129.91	ND
	CH(m)	128.26	129.03	128.18	128.56	129.23	129.20	ND
	CH(p)	125.33	126.12	125.29	125.68	126.28	126.29	ND

Table A7.3 ^{1}H NMR Data of Common Solvents and Contaminants

	Proton	Mult	THF-d_8	CD$_2$Cl$_2$	Toluene-d_8	C$_6$D$_5$Cl	TFE-d_3
Solvent residual peak			1.72 3.58	5.32	2.08 6.97 7.01 7.09	6.96 6.00 7.14	5.02 3.88
H$_2$O	OH	s	2.46	1.52	0.43	1.03	3.66
Acetic acid	CH$_3$	s	1.89	2.06	1.57	1.76	2.06
Acetone	CH$_3$	s	2.05	2.12	1.57	1.77	2.19
Acetonitrile	CH$_3$	s	1.95	1.97	0.69	1.21	1.95
Benzene	CH	s	7.31	7.35	7.12	7.20	7.36
tert-Butyl alcohol	CH$_3$	s	1.15	1.24	1.03	1.12	1.28
	OH	s	3.16			1.30	2.20
Chloroform	CH	s	7.89	7.32	6.10	6.74	7.33
18-Crown-6	CH$_2$	s	3.57	3.59	3.36	3.41	3.64
Cyclohexane	CH$_2$	s	1.44	1.44	1.40	1.37	1.47
1,2-Dichloroethane	CH$_2$	s	3.77	3.76	2.9	3.26	3.71
Dichloromethane	CH$_2$	s	5.51	5.33	4.32	4.77	5.24
Diethyl ether	CH$_3$	t	1.12	1.15	1.10	1.10	1.20
	CH$_2$	q	3.38	3.43	3.25	3.31	3.58
Diglyme	CH$_2$	m	3.43	3.57	3.43	3.49	3.67
	CH$_2$	m	3.53	3.50	3.31	3.37	3.62
	OCH$_3$	s	3.28	3.33	3.12	3.16	3.41
Dimethylformamide	CH	s	7.91	7.96	7.57	7.73	7.86
	CH$_3$	s	2.88	2.91	2.37	2.51	2.98
	CH$_3$	s	2.76	2.82	1.96	2.30	2.88
Dioxane	CH$_2$	s	3.56	3.65	3.33	3.45	3.76
Dimethoxyethane	CH$_2$	s	3.43	3.49	3.31	3.37	3.40

Continued

Table A7.3 ^{1}H NMR Data of Common Solvents and Contaminants—cont'd

	Proton	Mult	THF-d_8	CD$_2$Cl$_2$	Toluene-d_8	C$_6$D$_5$Cl	TFE-d_3
	OCH$_3$	s	3.28	3.34	3.12	3.17	3.61
Ethane	CH$_3$	s	0.85	0.85	0.81	0.79	0.85
Ethanol	CH$_3$	t	1.10	1.19	0.97	1.06	1.22
	CH$_2$	q	3.51	3.66	3.36	3.51	3.71
	OH	s	3.30	1.33	0.83	1.39	ND
Ethyl acetate	CH$_3$	s	1.94	2.00	1.69	1.78	2.03
	CH$_2$	q	4.04	4.08	3.87	3.96	4.14
	CH$_3$	t	1.19	1.23	0.94	1.04	1.26
Ethylene	CH$_2$	s	5.36	5.40	5.25	5.29	5.40
Ethylene glycol	CH$_2$	s	3.48	3.66	3.36	3.58	3.72
H Grease	CH$_3$	m	0.85–0.91	0.84–0.90	0.89–0.96	0.86–0.92	0.88–0.94
	CH$_2$	br s	1.29	1.27	1.33	1.30	1.33
Hexamethylbenzene	CH$_3$	s	2.18	2.20	2.10	2.10	2.24
n-Hexane	CH$_3$	t	0.89	0.89	0.88	0.85	0.91
	CH$_2$	m	1.29	1.27	1.22	1.19	1.31
Hexamethyldisiloxane	CH$_3$	s	0.07	0.07	0.10	0.10	0.08
HMPA	CH$_3$	d	2.58	2.60	2.42	2.47	2.63
Hydrogen	H$_2$	s	4.55	4.59	4.50	4.49	4.53
Imidazole	CH(2)	s	7.48	7.63	7.30	7.53	7.61
	CH(4,5)	s	6.94	7.07	6.86	7.01	7.03
Methane	CH$_4$	s	0.19	0.21	0.17	0.15	0.18
Methanol	CH$_3$	s	3.27	3.42	3.03	3.25	3.44
	OH	s	3.02, 1.09			1.30	ND
Nitromethane	CH$_3$	s	4.31	4.31	3.01	3.59	4.28
n-Pentane	CH$_3$	t	0.89	0.89	0.87	0.84	0.90
	CH$_2$	m	1.31	1.30	1.25	1.23	1.33
Propane	CH$_3$	t	0.90	0.90	0.89	0.84	0.90

	Group	Mult					
	CH2	sep	1.33	1.32	1.32	1.26	1.33
2-Propanol	CH3	d	1.08	1.17	0.95	1.04	1.20
	CH	sep	3.82	3.97	3.65	3.82	4.05
Propene	CH3	dt	1.69	1.71	1.50	1.58	1.70
	CH2(1)	dm	4.89	4.93	4.92	4.91	4.93
	CH2(2)	dm	4.99	5.03	4.98	4.98	5.03
	CH	m	5.79	5.84	5.70	5.72	5.87
Pyridine	CH(2)	m	8.54	8.59	8.47	8.51	8.45
	CH(3)	m	7.25	7.28	6.67	6.90	7.40
	CH(4)	m	7.65	7.68	6.99	7.25	7.82
Pyrrole	NH	br t	9.96	8.69	7.71	8.61	ND
	CH(2)	m	6.66	6.79	6.43	6.62	6.84
	CH(3)	m	6.02	6.19	6.27	6.27	6.24
Pyrrolidine	CH2(2)	m	2.75	2.82	2.54	2.64	3.11
	CH2(3)	m	1.59	1.67	1.36	1.43	1.93
Silicone grease	CH3	s	0.11	0.09	0.26	0.14	0.16
Tetrahydrofuran	CH2(2)	m	3.62	3.69	3.54	3.59	3.78
	CH2(3)	m	1.79	1.82	1.43	1.55	1.91
Toluene	CH3	s	2.31	2.34	2.11	2.16	2.33
	CH(2,4)	m	7.10	7.15	6.96–7.01	7.01–7.08	7.10–7.30
	CH(3)	m	7.19	7.24	7.09	7.10–7.17	7.10–7.30
Triethylamine	CH3	t	0.97	0.99	0.95	0.93	1.31
	CH2	q	2.46	2.48	2.39	2.39	3.12

Table A7.4 ^{1}H NMR Data of Common Solvents and Contaminants

	Proton	Mult	CDCl$_3$	(CD$_3$)$_2$CO	(CD$_3$)$_2$SO	C$_6$D$_6$	CD$_3$CN	CD$_3$OD	D$_2$O
18-Crown-6	CH$_2$	s	3.67	3.59	3.51	3.39	3.51	3.64	3.80
Ethane	CH$_3$	s	0.87	0.83	0.82	0.80	0.85	0.85	0.82
Ethylene	CH$_2$	s	5.40	5.38	5.41	5.25	5.41	5.39	5.44
Hexamethylbenzene	CH$_3$	s	2.24	2.17	2.14	2.13	2.19	2.19	ND
Hexamethyldisiloxane	CH$_3$	s	0.07	0.07	0.06	0.12	0.07	0.07	0.28
Hydrogen	H$_2$	s	4.62	4.54	4.61	4.47	4.57	4.56	ND
Imidazole	CH(2)	s	7.67	7.62	7.63	7.33	7.57	7.67	7.78
	CH(4,5)	s	7.10	7.04	7.01	6.90	7.01	7.05	7.14
Methane	CH$_4$	s	0.22	0.17	0.20	0.16	0.20	0.20	0.18
Propane	CH$_3$	t	0.90	0.88	0.87	0.86	0.90	0.91	0.88
	CH$_2$	sep	1.32	1.31	1.29	1.26	1.33	1.34	1.30
	CH$_3$	dt	1.73	1.68	1.68	1.55	1.70	1.70	1.70
Propene	CH$_2$(1)	dm	4.94	4.90	4.94	4.95	4.93	4.91	4.95
	CH$_2$(2)	dm	5.03	5.00	5.03	5.01	5.04	5.01	5.06
	CH	m	5.83	5.81	5.80	5.72	5.85	5.82	5.90

Table A7.5 ^{13}C NMR Data of Common Solvents and Contaminants

	Carbon	THF-d_8	CD$_2$Cl$_2$	Toluene-d_8	C$_6$D$_5$Cl	TFE-d_3
Solvent signals		67.21	53.84	137.48	134.19	61.5
		25.31		128.87	129.26	126.28
				127.96	128.25	
				125.31	125.96	
				20.43		
Acetic acid	CO	171.69	175.85	175.30	175.67	177.96
	CH$_3$	20.13	20.91	20.27	20.40	20.91
Acetone	CO	204.19	206.78	204.00	204.83	32.35
	CH$_3$	30.17	31.00	30.03	30.12	214.98
Acetonitrile	CN	116.79	116.92	115.76	115.93	118.95
	CH$_3$	0.45	2.03	0.03	0.63	1.00
Benzene	CH	128.84	128.68	128.57	128.38	129.84
tert-Butyl alcohol	COH	67.50	69.11	68.12	68.19	72.35
	CH$_3$	30.57	31.46	30.49	31.13	31.07
Carbon dioxide	C	125.69	125.26	124.86	126.08	126.92
Carbon disulfide	C	193.37	192.95	192.71	192.49	196.26
Carbon tetrachloride	C	96.89	96.52	96.57	96.38	97.74
Chloroform	CH	79.24	77.99	77.89	77.67	78.83
18-Crown-6	CH$_2$	71.34	70.47	70.86	70.55	70.80
Cyclohexane	CH$_2$	27.58	27.38	27.31	26.99	28.34
1,2-Dichloroethane	CH$_2$	44.64	44.35	43.40	43.60	45.28
Dichloromethane	CH$_2$	54.67	54.24	53.47	53.54	54.46
Diethyl ether	CH$_3$	15.49	15.44	15.47	15.35	15.33
	CH$_2$	66.14	66.11	65.94	65.79	67.55
Diglyme	OCH$_3$	58.72	58.95	58.62	58.42	59.40
	CH$_2$	71.17	70.70	70.92	70.56	73.05
	CH$_2$	72.72	72.25	72.39	72.07	71.33
Dimethylformamide	CO	161.96	162.57	161.93	162.01	166.01
	CH$_3$	35.65	36.56	35.22	35.45	37.76
	CH$_3$	30.70	31.39	30.64	30.71	30.96
Dioxane	CH$_2$	67.65	67.47	67.17	66.95	68.52
Dimethoxyethane	CH$_2$	72.58	72.24	72.25	58.31	59.52
	OCH$_3$	58.72	59.02	58.63	71.81	72.87
Ethane	CH$_3$	6.79	6.91	6.94	6.91	7.01
Ethanol	CH$_3$	18.90	18.69	18.78	18.55	18.11
	CH$_2$	57.60	58.57	57.81	57.63	59.68
Ethyl acetate	CH$_3$	20.45	21.15	20.46	20.50	21.18
	CO	170.32	171.24	170.02	170.20	175.55
	CH$_2$	60.30	60.63	60.08	60.06	62.70
	CH$_3$	14.37	14.37	14.23	14.07	14.36
Ethylene	CH$_2$	123.09	123.20	122.92	122.95	124.08
Ethylene glycol	CH$_2$	64.35	64.08	64.29	64.03	64.87
H Grease	CH$_2$	30.45	30.14	30.31	30.11	ND
Hexamethylbenzene	C	131.88	132.09	131.72	131.54	134.04

Continued

Table A7.5 ^{13}C **NMR Data of Common Solvents and Contaminants—cont'd**

	Carbon	THF-d_8	CD$_2$Cl$_2$	Toluene-d_8	C$_6$D$_5$Cl	TFE-d_3
	CH$_3$	16.71	16.93	16.84	16.68	17.04
n-Hexane	CH$_3$	14.22	14.28	14.34	14.18	14.63
	CH$_2$(2)	23.33	23.07	23.12	22.86	24.06
	CH$_2$(3)	32.34	32.01	32.06	31.77	33.17
Hexamethyldisiloxane	CH$_3$	1.83	1.96	1.99	1.92	2.09
HMPA	CH$_3$	36.89	36.99	36.80	36.64	37.21
Imidazole	CH(2)	135.72	135.76	135.57	135.50	136.58
	CH(4)	122.20	122.16	122.13	121.96	122.93
Methane	CH$_4$	−4.90	−4.33	−4.34	−4.33	−5.88
Methanol	CH$_3$	49.64	50.45	49.90	49.66	50.67
Nitromethane	CH$_3$	62.49	63.03	61.14	61.68	63.17
n-Pentane	CH$_3$	14.18	14.24	14.27	14.10	14.54
	CH$_2$(2)	23.00	22.77	22.79	22.54	23.75
	CH$_2$(3)	34.87	34.57	34.54	34.26	35.76
Propane	CH$_3$	16.60	16.63	16.65	16.56	16.93
	CH$_2$	16.82	16.63	16.63	16.48	17.46
2-Propanol	CH$_3$	25.70	25.43	25.24	25.14	25.21
	CH	66.14	64.67	64.12	64.18	66.69
Propene	CH$_3$	19.27	19.47	19.32	19.32	19.63
	CH$_2$	115.74	115.70	115.89	115.86	116.38
	CH	134.02	134.21	133.61	133.57	136.00
Pyridine	CH(2)	150.57	150.27	150.25	149.93	149.76
	CH(3)	124.08	124.06	123.46	123.49	126.27
	CH(4)	135.99	136.16	135.1	135.32	139.62
Pyrrole	CH(2)	118.03	117.93	117.61	117.65	119.61
	CH(3)	107.74	108.02	108.15	108.03	108.85
Pyrrolidine	CH$_2$(2)	45.82	47.02	47.12	46.75	47.43
	CH$_2$(3)	26.17	25.83	25.75	25.29	25.73
Silicone grease	CH$_3$	1.20	1.22	1.37	1.09	2.87
Tetrahydrofuran	CH$_2$(2)	68.03	68.16	67.75	67.64	69.53
	CH$_2$(3)	26.19	25.98	25.79	25.68	26.69
Toluene	CH$_3$	21.29	21.53	21.37	21.23	21.62
	C(1)	138.24	138.36	137.84	137.65	139.92
	CH(2)	129.47	129.35	129.33	129.12	130.58
	CH(3)	128.71	128.54	128.51	128.31	129.79
	CH(4)	125.84	125.62	125.66	125.43	126.82
Triethylamine	CH$_3$	12.51	12.12	12.39	11.87	9.51
	CH$_2$	47.18	46.75	46.82	46.36	48.45

Table A7.6 ^{13}C NMR Data of Common Solvents and Contaminants

	Carbon	CDCl$_3$	(CD$_3$)$_2$CO	(CD$_3$)$_2$SO	C$_6$D$_6$	CD$_3$CN	CD$_3$OD	D$_2$O
Carbon dioxide	C	124.99	125.81	124.21	124.76	125.89	126.31	ND
Carbon disulfide	C	192.83	193.58	192.63	192.69	193.60	193.82	197.25
Carbon tetrachloride	C	96.34	96.65	95.44	96.44	96.68	97.21	96.73
18-Crown-6	CH$_2$	70.55	71.25	69.85	70.59	71.22	71.47	70.14
Ethylene	CH$_2$	123.13	123.47	123.52	122.96	123.69	123.46	ND
Hexamethylbenzene	C	132.21	132.22	131.10	131.79	132.61	132.53	ND
	CH$_3$	16.98	16.86	16.60	16.95	16.94	16.90	ND
Hexamethyldisiloxane	CH$_3$	1.97	2.01	1.96	2.05	2.07	1.99	2.31
Imidazole	CH(2)	135.38	135.89	135.15	135.76	136.33	136.31	136.65
	CH(4)	122.00	121.96	122.31	122.16	122.78	122.60	122.43
Methane	CH$_4$	−4.63	−5.33	−4.01	−4.29	−4.61	−4.90	ND
Propene	CH$_3$	16.63	19.42	19.20	19.38	19.48	19.50	ND
	CH$_2$	16.37	116.03	116.07	115.92	116.12	116.04	ND
	CH	25.14	134.34	133.55	133.69	134.78	134.61	ND
Pyrrole	CH(2)	117.77	117.98	117.32	117.78	118.47	118.28	119.06
	CH(3)	107.98	108.04	107.07	108.21	108.31	108.11	107.83
Pyrrolidine	CH$_2$(2)	46.93	ND	46.51	46.86	47.57	47.23	46.83
	CH$_2$(3)	25.56	ND	25.26	25.65	26.34	26.29	25.86

Reprinted with Permission from Fullmer, Miller, Sherden, Gottlieb, Nudelman, Stoltz, Bercaw, Goldberg, 2010. Copyright 2010, American Chemical Society.

Appendix 8
Acidities of Organic Functional Groups

These acidities are expressed as pKa compared to water. Those in Fig. A8.1A were determined in aqueous media, and those in Fig. A8.1B were determined by extrapolation from aqueous media. The hydrogen to which the pKa refers is explicitly shown. These data came from several sources (House, 1972; March, 1992; Reichardt and Welton, 2010).

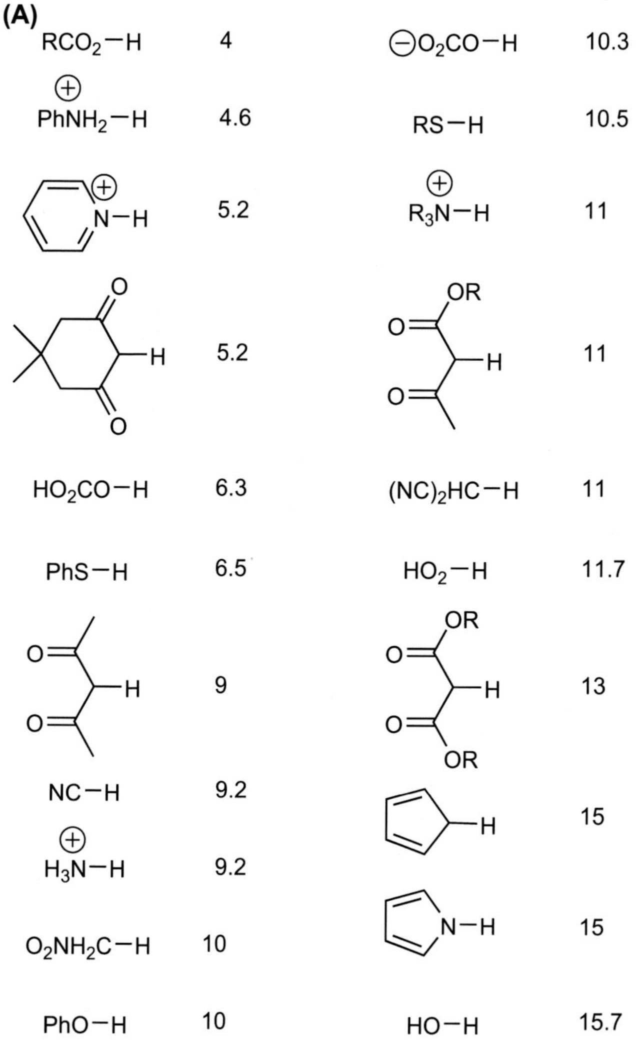

(A)

Structure	pKa	Structure	pKa
RCO_2–H	4	$^{\ominus}O_2CO$–H	10.3
$PhNH_2^{\oplus}$–H	4.6	RS–H	10.5
pyridinium N–H	5.2	$R_3N^{\oplus}$–H	11
dimedone –H	5.2	ester/ketone OR –H	11
HO_2CO–H	6.3	$(NC)_2HC$–H	11
PhS–H	6.5	HO_2–H	11.7
diketone –H	9	ester OR –H	13
NC–H	9.2	cyclopentadiene OR –H	15
$H_3N^{\oplus}$–H	9.2	pyrrole N–H	15
O_2NH_2C–H	10	HO–H	15.7
PhO–H	10		

Figure A8.1 Acidities of organic functional groups (as expressed in pKa) compared to water. (A) Aqueous media and (B) extrapolation from aqueous media.

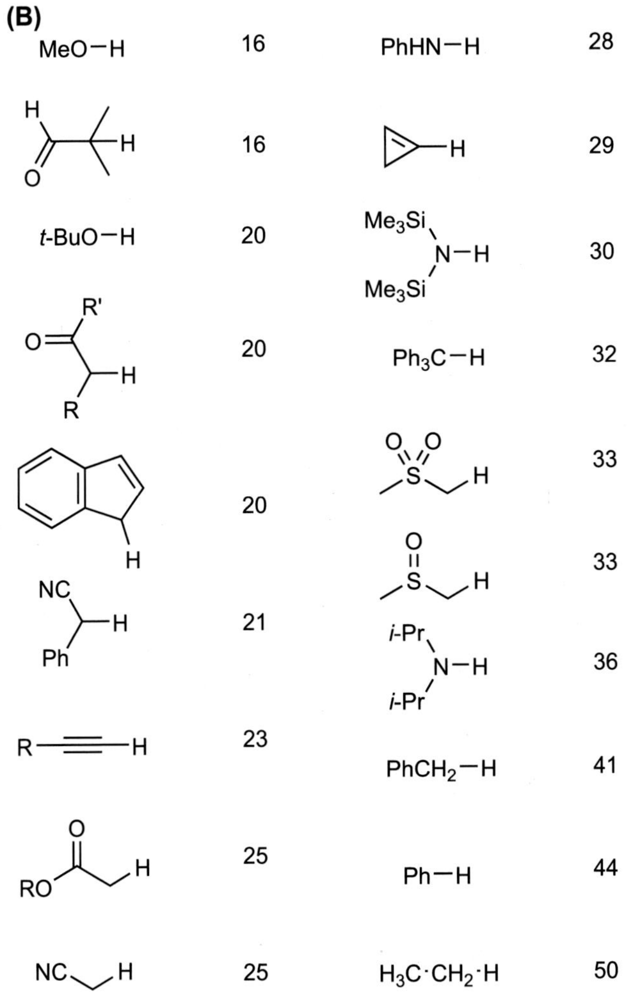

(B)

MeO–H	16	PhHN–H	28
(acetaldehyde/isopropyl structure)	16	(cyclopropene)–H	29
t-BuO–H	20	Me₃Si–N(H)–SiMe₃	30
O=C(R')–CHR–H	20	Ph₃C–H	32
(indene structure)	20	CH₃–S(=O)₂–CH₂–H	33
NC–CH(Ph)–H	21	CH₃–S(=O)–CH₂–H	33
R≡≡–H	23	i-Pr–N(H)–i-Pr	36
RO–C(=O)–CH₂–H	25	PhCH₂–H	41
NC–CH₂–H	25	Ph–H	44
		H₃C·CH₂·H	50

Figure A8.1 Continued.

References

House, H.O., 1972. Modern Synthetic Reactions, second ed. Benjamin, Menlo Park.

March, J., 1992. Advanced Organic Chemistry: Reactions, Mechanisms, and Structure, fourth ed. Wiley, New York.

Reichardt, C., Welton, T., 2010. Solvents and Solvent Effects in Organic Chemistry, fourth ed. Wiley-VCH, Weinheim.

Appendix 9
Acidities of Organic Functional Groups in DMSO

Acidities of organic functional groups determined in dimethyl sulfoxide (Fig. A9.1) (Bordwell, 1988). The hydrogen to which the pKa refers is explicitly shown.

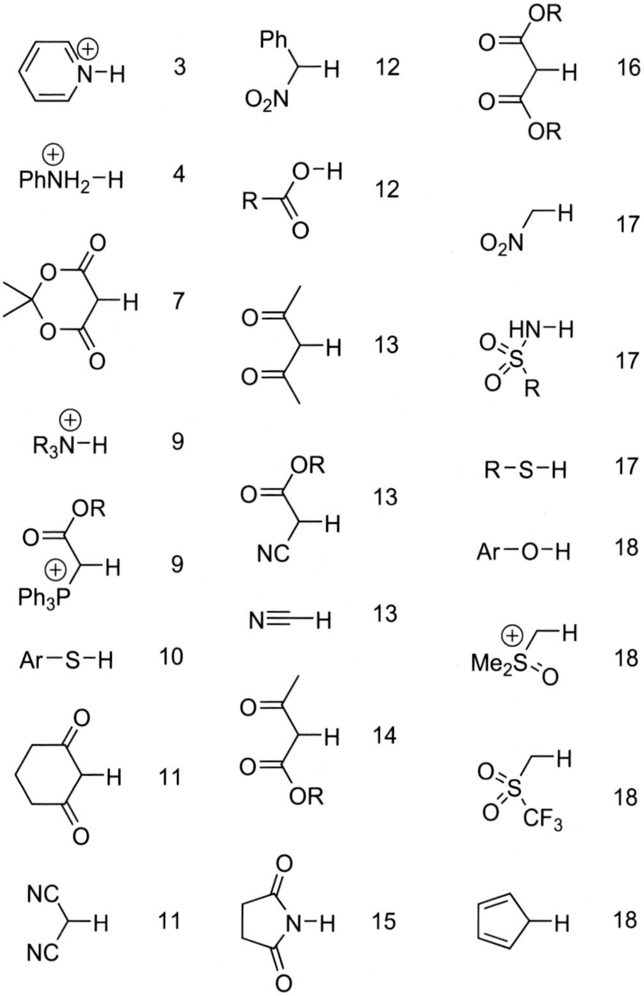

Figure A9.1 Acidities (as expressed in pKa) of organic functional groups in dimethyl sulfoxide.

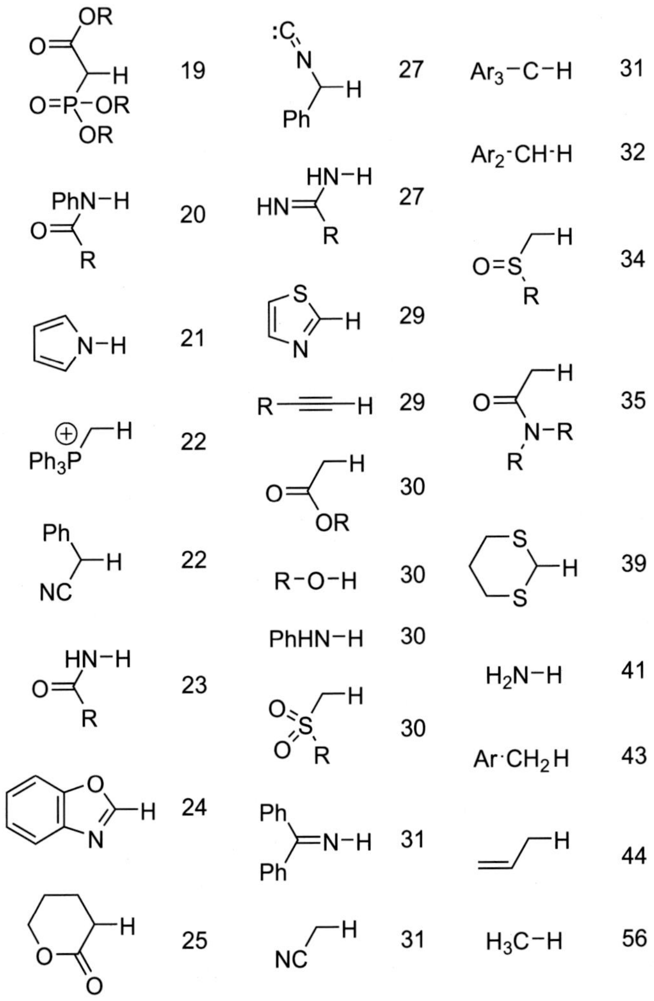

Figure A9.1 Continued.

Reference

Bordwell, F.G., 1988. Equilibrium acidities in dimethyl sulfoxide solution. Acc. Chem. Res. 21, 456–463. http://dx.doi.org/10.1021/ar00156a004.

Appendix 10
Stuck Joints

Frozen ground-glass joints are a common problem. To keep joints from freezing, grease the joint every time it is assembled. Nevertheless, even with good greasing, joints can still seize. Grease may be leached out by the solvent, especially if it is heated. Solids stuck in the joint (e.g., salts from a reaction) can also cause a joint to seize. This is one reason you might want to consider Teflon tape or a Teflon sleeve as an alternative if you anticipate this sort of problem. Several strategies to free stuck joints are available, but creativity is typically also essential to solve such situations.

SAFETY NOTE

Trying to free stuck glass joints can result in serious bodily injury or death if the glass breaks in the process. Wear eye protection and sturdy work gloves and wrap the glass in a heavy lab towel to lower the potential for serious cuts or injury. Do not use heat or flame if there is any potential for hazardous or flammable residues or vapors to ignite. Never heat a closed system. If you are not comfortable with assessing or understanding the risk involved, do not attempt freeing joints yourself.

The following methods can be conducted once the flask has been emptied (through a sidearm or another opening), cleaned, and dried to ensure that there are no harmful residues or solvents present:

1. Soak the joint overnight, either in a base bath (KOH/isopropanol) or an ultrasonicator bath. Look for uniform wetting of the ground joint surfaces, which indicates liquid penetration.
2. Put a solvent, glycerol, or a penetrating oil on the outside of the joint and try to work it down into the joint as the parts are rocked. Again, look for a "clear" joint.
3. Immerse the joint in a container of freshly poured carbonated water (soda water). You should be able to see the liquid penetrate between the ground surfaces. Remove the joint and rinse once the surfaces are wet (allow 5–10 min). After ensuring that half or more of the inner surface is wet, gently warm the wall of the outer joint by rotating it for 15–20 s over a low flame. Remove from the flame and, taking care to protect hands/fingers, gently twist the two members apart. Repeat from the beginning if required.
4. Apply heat to the outer glass joint so that it expands, freeing it from the inner joint. For this method to work well, it is essential that the outer joint expands rapidly without similar expansion of the inner joint. Because their heating rates are too slow to promote differential expansion, heat guns or Bunsen burners typically do not work well in this method. Heat the joint with a glassblowing torch in a cool to moderate (yellow) flame with just a touch of oxygen. Direct the flame to any adhesion points that are visible, but do not hold the flame on any one spot for too long—you will lose the effect of differential expansion. After a few seconds of rotation of the joint in the flame, use an insulated, gloved hand or a hook to try to separate the parts. Glassware must be annealed if a torch has been used on the joint.
5. Put the apparatus through an annealing cycle in the glass shop's annealing oven.

If a glass stopper has become stuck in a joint, hold the head of the stopper with a gloved hand or a clamp. Place a wooden stick or dowel against the outer joint and tap lightly but sharply on it with a hammer. Assistance to hold one of the items or to catch the parts as they separate can be helpful with this technique.

If this does not work, put a lab towel over the head of the stopper and loosely place a crescent (adjustable) wrench on the head. Do not let the metal touch the head. Try to rotate the stopper with the wrench, but do not apply so much force that you break the stopper (which, unfortunately, is easy to do). If the stopper breaks and it is hollow, smash out its remains. If it is solid, move on to other options.

When a separatory funnel with Teflon stopcock is used, the stopcock should not be overtightened because it will warp, and should not be exposed to sudden heat, since Teflon expands much faster than glass. Conversely, stuck Teflon stopcocks can be freed by immersing them in dry ice/acetone.

If none of these methods work, ask a glassblower, who may have access to specialized tools such as glass joint or stopper pullers (similar to gear pullers, for you mechanics).

Index

Diisopropyl ether, 67–68
Dimethoxyethane, 71
p-Dimethylaminobenzaldehyde, 244
Dimethylsulfoxide, 66–67, 71
  acidity, 86
  deuterated, 126
  reaction solvent, 65
Dinitrophenylhydrazine, 221, 243
Dioctyl phthalate, 46–47
Dioxanes, 205
Dispersions, 83–84
Distillation
  head, 18
  of reagents, 18
  at reduced pressure, 18
  short path, 18–20
  of solvents, 67
Doty susceptibility matching plug set, 166,
  166f
DRIERITE, 43
Dry ice
  coolant, 51–53, 61–62, 73, 145
  reactant, 129–130
Drying agents, 69–70
Drying oven, 41, 194
Drying pistol, 172
Drying tower, 43, 75

E

Ehrlich's reagent, 244
Electronic laboratory notebooks (ELNs),
  80–82, 81f
Electronic resources, 14–15
Electronic timer, 111
Electrons, solvated, 74–75
Emulsion, 130–131
Enolate, 84–85
Ethanol
  aqueous, 169–170
  quench, 54
  solvent, 170–171
  stabilizer, 73–74
Ethyl acetate
  extraction, 129
  quench, 90
  solvent, 227–228
Ethylene glycol, 84
Ethylene oxide, 20–21

Ethyl isopropyl ketone, 84–85
Ethyl levulinate, 90–91
Ethyl vinyl ether, 186
Evaporation
  centrifugal evaporators, 141, 141f
  heat gun, 141, 142f
  rotovap bulb prevents bumps, 139–140,
    140f
Evaporator, centrifugal, 84
Evaporator, rotary, 140f, 141
Experimental descriptions, 185
Explosion, 2, 25, 27, 94b, 192–193,
  220–225

F

Ferric chloride, 244
Filter
  adapter, 142
  glass frit, 182–183
  paper, 118
  Schlenk-ware frit, 106
Filtering bell jar, 112–113, 113f
Flame ionization detector (FID), 122–123
Flash chromatography, 78–80, 153–157
Flash point, 56–57, 191
Flask
  Erlenmeyer, 38, 105, 130–131
  pear-shape, 48, 139–140
  round-bottom, 23, 38, 151–153
  tip, 51
Floods, 51–53
Florisil, 158–159
Flow indicator, 51–53
Flushing oil, 145
Forceps, 74
Fragmentation pattern, 172
Freezing point, 28, 61–62, 90, 237t
Frontier, 35
Fume hood, 32–33, 192, 239
Functional groups
  acidities, 86
  spectroscopy, 169
  staining, 119
Funnel, addition, 220

G

Gas chromatography (GC), 122–123
General TLC stains, 243

Sensitizer
  chemical, 54
  triplet, 100. *See also* Triplet sensitizer
Separatory funnel, 105, 264
Septum, 21–22, 60, 89, 218
Serum cap, 24f, 47–48
Serum stopper, 41, 47–48
Short-path distillation
  cow adapter, 19f
  short-path still head, 18f
  vapor pressure nomograph, 20f
  Vigreux column, 19f
Sigma-Aldrich, 11
Silanization, 38
Silica gel
  chromatography, 153
  filtration, 157
  HPLC, 124
  TLC, 78–80
Silver nitrate, 158, 206–207
Singlet oxygen, 100
Small gas tank, 29, 30f
SMILES, 11
Sodium, 196, 210
Sodium bicarbonate, 196
Sodium bisulfite, 99–100, 129
Sodium borohydride, 221–222
Sodium carbonate, 196
Sodium chloride, 130–131
Sodium cyanoborohydride, 136
Sodium fluoride, 134
Sodium hydride, 221–222
Sodium hydroxide, 25–26, 74–75, 198
Sodium sulfate, 78–80
Sodium thiosulphate. *See* $Na_2S_2O_3$
Solvent delivery systems (SDSs), 71–72
Solvent partitioning, 127–128, 235
Solvent properties, 65, 131t
Solvent reservoir, 155
Solvent still, 57, 69
Soxhlet extractor, 93, 93f
Sparging, 24, 73f, 231–232
Specialized TLC stains, 243–244
Specific chemical safety training, 2
Starch, 135–136
Starting material, 11–13, 17, 35, 111, 142
Static electricity, 193
Statistical methods, 65, 188
Stereochemistry, 15, 85–86

Stille coupling, 134
Stoichiometry, 81–82, 84–87
Stopcock, 25, 43, 90, 143, 264
Stopcock grease, 40, 127
Stoppers
  glass, 264
  rubber, 47, 98, 141
Sulfides, 137
Sun lamp, 100
Supporting material, 14
Sure/Seal, 21–22, 220
Surface tension, 130–131, 166–167
Susceptibility plug, 166
Suzuki coupling, 133
Syringes
  cleaning, 90
  gastight, 87–88, 88f, 102
  Luer, 87–88, 87f, 102
  microliter, 87–88, 90
  multifit, 87–88, 87f
  plastic, 88, 99–100
  pump, 90
  tuberculin, 86–88

T

Tanks, 32–33
Tank valve, 29, 32–33, 74
Teflon, 37–38, 55, 264
Teflon sleeve, 40, 263
Tetrahydrofuran
  peroxides, 67–68
  quench, 183
  solvent, 47–48, 105
  with stoppers, 48
Tetrahydronaphthalene, 205
Tetramethylsilane, 40, 165
Tetraphenylporphyrin, 100
Theoretical plate, 18
Thermometer, 36–37, 111, 186
Thermometer adapter, 55
Thermostat, 62, 103
THERM-O-WATCH monitors, 56–57, 58f
Thimble, paper, 93
Thin film, 139, 169–170
Thin layer chromatography (TLC),
        153–154, 243–244
  chambers, 112–113, 118
  dips, 119–120, 121t

Printed in the United States
By Bookmasters